Explore, Navigate, Protect: The Diverse Uses of Radar

Mathew

Contents

Chapter 1

Introduction

Radar, short for RAdio Detection And Ranging, is defined by the IEEE Std. 686 as "An electromagnetic system for the detection and location of objects that operates by transmitting electromagnetic signals, receiving echoes from objects (targets) within its volume of coverage, and extracting location and other information from the echo signal" [1].

Radar has many applications, including but not limited to:

- Target detection

- Range, height, angle and velocity measurements

- Applications in guidance systems

- Weather monitoring

- Imaging

Inggs et al. offered a taxonomy of Electromagnetic (EM) sensors in [2] where they described two main categories of EM sensors, namely active and receive only. The authors defined active sensors as those with a dedicated radar transmitter that radiates EM energy in accordance with the defined radar application. The second category is that of 'receive only' systems, commonly referred to as 'passive radar'.

Passive radars use transmitters of opportunity where the radar operator has no influence on the waveforms. There are many examples of such systems in literature e.g. [3–9] to list a few. There are many names for 'passive', receive only radar found throughout literature such as Commensal Radar (CR), Passive Coherent Location (PCL) radar, Passive Bistatic Radar (PBR) and Passive Radar (PR). This **book** will use the most common term, PR, to describe such systems because this term is widely used and understood. In these forms of passive radar the operator has no influence on the transmitter, nor the transmitted waveform.

1.1 Overview of Passive Radar

Significant research into PR has been performed over the last two decades across various industries as PR technology has matured. There are several challenges unique to PR that include waveform suitability, performance complications owing to bistatic geometry and the effects of Direct/Multi-path Signal Interference (DSI) in the surveillance channel.

Target echoes are typically multiple orders of magnitude weaker than the DSI. Multipath versions of the direct signal can also be deleterious to PR performance. This multipath interference along the zero-Doppler ridge together with DSI and reflections from stationary scatterers can often mask target echoes. The problem of DSI associated with continuous-wave PR necessitates receivers with high dynamic range [4, 10–12].

As mentioned, PR is a type of radar system that utilises existing transmitter infrastructure to perform target detections. These uncooperative transmitters are referred to as illuminators of opportunity [10, 13]. Common illuminators of opportunity include analogue based systems such as High Frequency (HF) and FM radio as well as digital based systems such as Digital Audio Broadcast (DAB), Digital Video Broadcast Terrestrial (DVB-T), Digital Video Broadcast Terrestrial 2 (DVB-T2), Global Positioning System (GPS), Global System for Mobile communications (GSM), Digital Video Broadcast Satellite (DVB-S) and wireless local area networks (WiFi). While there are many different transmitters, not all of them are necessarily desirable.

In order for an illuminator of opportunity to be considered 'desirable', two main conditions need to be met [10, 14]:

1. The transmit power and antenna beam pattern must be sufficient for the desired coverage.

2. The modulation bandwidth of the illuminating signal should be sufficient to meet the desired range and Doppler resolutions.

3. Appropriate line of sight between potential transmitter-target and target-receiver pairs.

Unlike typical active monostatic radars, where the transmitter and receiver are co-located as shown in Figure 1.1, passive radars are inherently bistatic - meaning that there is dislocation between the transmitter and the receiver, as shown in Figure 1.2. As a result, the variety of channel combinations is huge, where waveform and spatial diversity can be exploited.

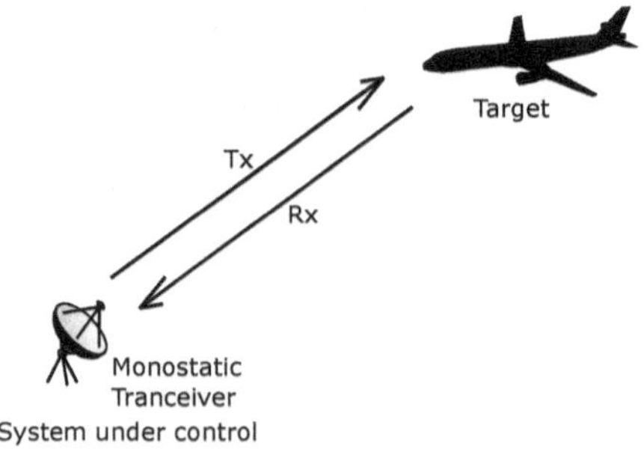

Figure 1.1: Basic monostatic radar geometry.

It can be seen from Figure 1.1 that the basic active monostatic radar consists of a Transmit (Tx)/Receiver (Rx) module that transmits a user-specified pulse and receives the resultant target echo. Both the transmitter and receiver chain are under full control of the radar operator. Contrasting this to the bistatic case of a PR, the fundamental difference is the baseline separation between the Tx and Rx sites as well as the lack of control over the transmitter infrastructure.

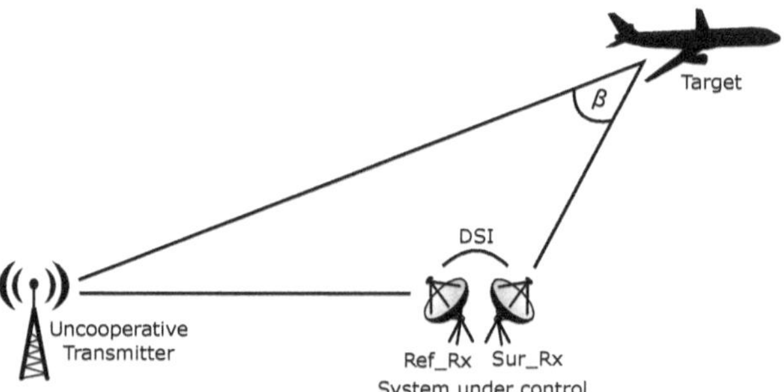

Figure 1.2: Typical bistatic passive radar system geometry, illustrating target detections utilising existing, uncooperative transmitter infrastructure.

Due to the lack of waveform control with a PR, the processing chain is normally different to that of an active radar. Most PR is Continuous Wave (CW) and therefore requires continuous monitoring of the transmitted waveform. To constantly monitor, a basic PR makes use of two receiver channels. One of the channels, referred to as the reference channel, is used to record the signal arriving at the receiver along the direct path form the transmitter. A second channel, referred to as the surveillance channel, is used to record signals within the designated surveillance coverage volume. Both the reference and surveillance channels are continuously receiving and digitising received signals.

One of the major advantages of passive radar is that it does not emit any EM energy, making it difficult to detect using Electronic Support (ES) systems and therefore more difficult to counter by conventional means such as anti-radiation homing. PR vulnerabilities against Electronic Countermeasures (ECM), however, have so far received little attention in open literature.

One of the perceived strengths of PR is its ability to be diverse in either frequency of operation, spacial coordinates or both. While the combination of spacial and/or frequency diversity of a PR is immense, their performance and tolerance against ECM is critically important along with their ability to provide potential Electronic Counter-countermeasures (ECCM).

1.1.1 Co-located and Separated Receiver Channels

The reference and surveillance channels are generally configured as two independent antennas in two main categories. The first category is referred to as a co-located reference and surveillance system, by far the most common, is depicted in Figure 1.3 while the second category is referred to as a separated reference and surveillance system, as shown in Figure 1.4.

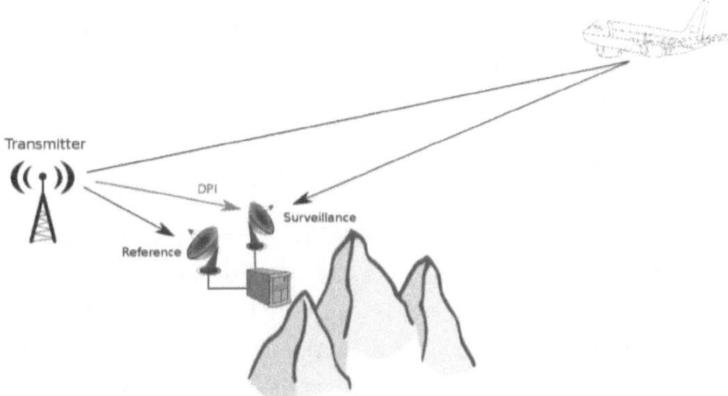

Figure 1.3: Co-located reference and surveillance system where both the reference and surveillance antennas are at the same physical location [11].

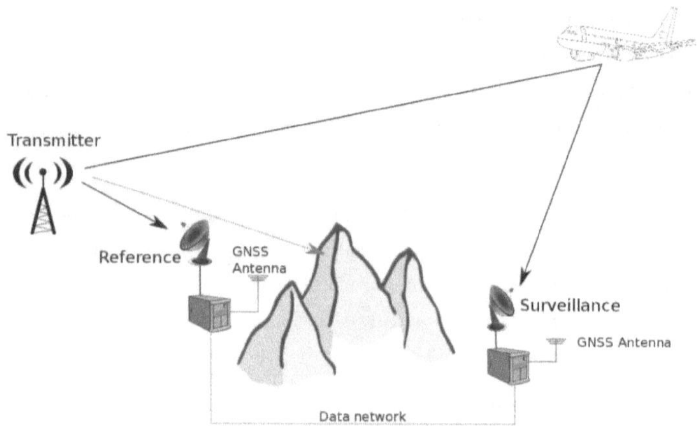

Figure 1.4: Separated reference and surveillance system where the reference and surveillance channels are at different locations, usually with physical shielding such as a large structure separating the two channels for additional DSI suppression [11].

The major benefit to utilising a separated reference and surveillance configuration is that the two receive channels can be spatially separated by a physical structure, leading to greatly reduced DSI. The most obvious drawback to such a configuration is the need for coherent synchronisation of the two channels. This is not the case with a co-located configuration however, co-located would typically systems experience increased levels DSI associated with being in direct Line-of-Sight (LoS) to the transmitter, as shown in Figure 1.3.

Both configurations require largely the same processing with the separated reference and surveillance configuration requiring additional data synchronisation and transportation steps as illustrated by the Global Navigation Satellite System (GNSS) antennas and data link network in Figure 1.4.

1.1.2 The Growing Need for Passive Radar

As mobile technology improves and becomes more integrated into our day-to-day lives, the drive to free up existing spectrum for the purposes of telecommunications and broadcast systems is at an all time high. Currently the greatest portion of spectrum is allocated for telecommunication services, with a much smaller allocation for radar, including Air Traffic Control (ATC) and navigation radar [15, 16].

In a 2014 report by the GSM association, found in [17], the authors found that between 1991 and 2008 50 billion USD had been raised through spectrum auctions, with this value expected to grow exponentially over the coming years. In the same report it was estimated that the annual cost of poor spectrum utilisation in India is in excess of 3.6 billion USD.

Combining the need to free up existing spectrum with the exponential growth of air traffic across the world, the prospect of a PR based Air Traffic Management (ATM) system becomes evermore appealing. In order for PR based ATM systems to be considered in a commercial sense, it needs to be reliable and potential vulnerabilities to ECM fully understood [18–20].

1.2 Overview of Electronic Attacks

To investigate the performance of PR in the presence of ECM, it is important to discuss the common forms of ECM applied to radar systems. There are various types of jamming techniques used to counter radar systems, including [21]:

- Stand-off jamming

- Stand-in jamming

- Self protection and escort jamming

This can be further broken down into basic sub-categories such as:

- Broadband or barrage jamming

- Narrowband or spot jamming

- Deception jamming using DRFM

- Deception jamming using decoys

1.2.1 Barrage vs. Spot Jamming

As the name suggests, barrage jamming attempts to radiate EM energy to cover a very wide bandwidth while spot jamming is very narrowband in comparison. One of the benefits to barrage jamming is that the user requires very little knowledge of the enemy system as all frequencies within the jammers band get attacked. The downside to this is that the total jammer energy is spread across a very wide band, leading to much higher power levels being required. This also leads to barrage jamming being easier to detect than spot jamming.

Spot jamming, on the other hand, greatly improves the jammers performance as almost all of the energy is transmitted in the direction of the threat radar. This narrow band technique allows the jammer operator to operate the jammer at a much lower power level in order to achieve the same results as with a broadband jammer. The downside to spot jammers, however, is that they require detailed knowledge of the threat-radar frequency and bandwidth.

1.2.2   Stand-off vs. Stand-in Jamming

Stand-off jamming, as the name suggests, is a technique where the jammer is placed at a safe distance from the radar, this distance is equal to the radars instrumented range. This allows masking of friendly targets while remaining out of harms way.

Contrasting this with stand-in jamming where the jammer is placed in a considerably more vulnerable position within range for an attack such as by an anti-radiation missile. The major benefit to using stand-in jamming as opposed to stand-off jamming is the reduction in the power requirements. In both cases, the jammer operates as an independent system that is placed in a strategic location with the goal of masking a target within the radar surveillance volume.

1.2.3   Self Protection and Escort Jamming

Self protection jamming refers to a case where a target carries a jammer while escort jamming is when the target is escorted by friendly craft that jam threat radars. The major downside to escort jamming is that the target or target escort becomes a beacon for ES systems as it emits EM energy. Another limitation of self protection and escort jamming is the increase in crossover and burn-through range.

The crossover range is the range at which the Jammer-to-Signal ratio (JSR) becomes ineffective. The burn-through range is the range at which the jamming signal equals the target echo and the jammer becomes ineffective. This occurs because the target approaches the radar and its echo strength increases by a factor of R^4 while the jamming signal increases at the receiver by a factor of R^2.

1.2.4 Sidelobe vs. Mainlobe Jamming

Another aspect that must be considered when jamming a radar is whether the transmitted energy is aimed at the mainlobe or the sidelobe of the radar. A common technique used to counter jamming is to steer a null of the antenna into the direction of the jamming source. This requires a well calibrated ES receiver with direction finding capability in order to detect the presence and location of the jamming source.

1.2.5 Main Jamming Techniques Investigated

Broadband noise-jamming

One of the key metrics to assess the performance of a jammer against a radar is the JSR at the receiver. The aim of noise jamming is to raise the noise floor of the PR such that targets are masked. The minimum effective JSR is determined by the radar sensor and is often related to the integration gain achieved by the processing chain. In the case of a PR, with the target echo potentially 90 dB below the DSI, even a moderate JSR could raise the noise floor enough to mask the target.

The advantage of this noise jamming technique is that it covers the PR range and Doppler extent. The disadvantage however, is that the noise power is widely spread, so a jammer with higher ERP would be needed if the jamming signal is to be effective.

Jamming Through Coherent Integration

Unlike noise jamming that simply raises the receiver noise floor, coherent integration is achieved by transmitting a part of the radar waveform back in such a way that it causes additive or coherent integration when passed through the radar processing chain. This is achieved in two ways, either through attacking the deterministic components of the illuminator signal or through the use of Digital Radio Frequency Memory (DRFM) based systems. It must be noted however, that most repetitive signals found in both surveillance and reference channels, not exhibiting Doppler, will likely be removed by the cancellation processing.

In contrast to noise jamming, coherent jamming requires a lower jammer ERP to generate false targets. False targets can therefore be placed in the receiver using relatively low power and while it is a strength of the tracking process to remove false targets, a system that is able to overload the tracker with a large number of false targets might have success.

1.3 Problem Statement

Before PR can be used in a military context, the system vulnerabilities to ECM need to be properly understood. As will be shown in Chapter 2.2, there is clearly a gap in the literature with regards to ECM applied to PR with the only relatively comprehensive open research performed by Inggs et al., found in [2]. This is due in part to the classified nature of the research as well as the assumed resilience PR has to ECM. Another contributing factor to the lack of sufficient open literature is due to the vast variety of PR architectures, making it difficult to comprehensively and quantitatively draw conclusions as to what the real world effects will be with regards to ECM and ECCM.

This **book** aims to fill the important gap in open literature regarding ECM and ECCM in the PR context. The objectives for this research are therefore:

- Perform a complete and comprehensive investigation on the effectiveness of various countermeasures, focusing specifically on ECM applied to FM and DVB-T2 based PR.

- Propose basic ECCM to mitigate the effects of various Electronic Attacks (EA) against both FM and DVB-T2 PR.

- Provide representative experimental results from a real FM PR to verify the simulated results.

1.3.1 Research Novelty

The goal of this **book** is to provide the first comprehensive, open investigative and quantitative study on ECM and ECCM applied to FM and DVB-T2 based PR. The novel aspects of this research include:

FM Passive Radar

- The effect of the DSI canceller in FM PR in the presence of jamming is quantified.

- A complete waveform study is performed to demonstrate the optimal FM jamming waveform.

- Basic ECCM is presented to counter potential jamming of FM PR.

- A representative measurement of a real FM PR is shown to validate the simulated results

DVB-T2 Passive Radar

- An in depth review of the performance of the two most common processing techniques, 'mismatched filtering' and 'inverse filtering', in the presence of noise jamming is presented.

- The deterministic components of the DVB-T2 waveform are shown to be an effective form of ECM for both mismatched filtering and inverse filtering processing techniques.

- Basic ECCM is presented to counter potential pilot attacks on DVB-T2 PR.

1.4   book Outline

The remainder of this **book** is divided into the following chapters:

A comprehensive review of PR literature is provided in Chapter 2. The chapter begins with a comprehensive overview of research performed over many years in the field of PR before providing a brief but comprehensive review of available literature focusing specifically on ECM applied to PR.

FM PR is covered in Chapter 3. The chapter begins by providing a comprehensive review of the FM waveform before detailing the typical FM PR processing chain that is used in this **book**. The simulation set-up is then discussed before the effects of DSI cancellation in the presence of jamming is analysed.

Chapter 4 presents a comprehensive jammer waveform analysis followed by an exposition of the measured results in Chapter 5 where the simulated results are compared to a real world FM PR.

Chapter 6 begins by providing a complete overview of the DVB-T2 signal structure. The typical processing methods, which include mismatched and inverse filtering are then presented and the Ambiguity Function (AF) is discussed.

Chapter 7 investigates the effectiveness of different types EA applied to DVB-T2 PR. Each attack is demonstrated and compared using both mismatched and inverse filtering processing methods before a brief section on potential ECCM is provided. The chapter then concludes with a short discussion on results of each simulation.

The **book** concludes with Chapter 8 where conclusions and future work is discussed.

A brief overview of required jamming power levels for different scenarios is given in Appendix A that demonstrates the importance of having intelligence regarding the location of the enemy PR receiver when attempting to jam it.

Chapter 2

Literature Critique

This chapter critiques the relevant literature on PR and assesses current research programmes across various research groups. Various commercial systems are discussed prior to providing a review of literature focusing specifically on ECM applied in the PR context where a clear gap in open literature is shown. The chapter concludes with a summary of the current state-of-the-art pertaining to PR countermeasures.

2.1 Passive Radar Research

This section provides a review of some of the important literature related to PR. Beginning with a discussion of the numerous illuminators of opportunity such as HF radio signals, Very High Frequency (VHF) FM radio signals, DAB, DVB-T, DVB-T2, GSM, GPS, and satellite broadcast services such as DVB-S. Early work on PR is then discussed before moving to system limitations and suggested solutions on overcoming these limitations.

2.1.1 Early Work in Passive Radar

While research in the field of PR has increased over the last decade, the concept of PR is as old as radar itself [22–24]. In a 2005 paper by Howland [22], the authors describe how the famous Daventry experiment in 1935 was the first PR demonstration as it utilised a British Broadcasting Corporation (BBC) broadcast illuminator of opportunity rather than its own dedicated transmitter. Other examples of early passive radar systems were developed in World War II (WWII) by the German armed forces to exploit transmissions from the British Chain Home radars [23]. After WWII, little interest in passive radar remained due technical limitations at the time as well as the emergence of monostatic active radar systems until about the late 1980s [23]. This renewed interest was sparked by improvements in technology, most notably the availability of cheap computing resources such as General Purpose Graphics Processing Units (GP-GPUs), high resolution Analogue-to-Digital Converters (ADC) [11, 25] and high performance Software Defined Radio (SDR) such as the Ettus SDR platform [26].

In work by Griffiths and Long [3], analogue Television (TV) broadcasts were used from the Crystal Palace transmitter in South London as the illuminator of opportunity to detect aircraft landing and taking off from Heathrow airport. In these experiments, the receiver was placed 11.8 km away from the transmitter which had an omni-directional antenna with an ERP of 1 MW. The system operated on TV channels between 487.25 and 567.25 MHz.

Each channel had a bandwidth of 8 MHz, resulting in an ERP of 250 kW/channel. It was found that the autocorrelation function of the sync-plus-white waveform of the analogue TV signal exhibited high sidelobes with poor range resolutions and severe ambiguities at 9 600 m and integer multiples thereof. It was therefore concluded that the analogue TV signal was not a good illuminator of opportunity due to the sync pulses. At the time, ADC technology was limited to 8 bits or 48 dB of dynamic range which proved to be insufficient with Griffiths noting that the processing gain can not exceed the dynamic range of the system. With 8 bits, and likely even lower Effective Number of Bits (ENOB), this meant that a maximum of 48 dB of potential integration gain was likely insufficient for most targets of interest [10].

In 1992 and revisited a decade later in 2002, Griffiths et al. [27, 28] discussed the use of space-borne Synthetic Aperture Radar (SAR) as an illuminator of opportunity for a ground based PR. This has been further explored in recent years with satellite based illuminators being used to conduct Inverse Synthetic Aperture Radar (ISAR) experiments for imaging and coastal surveillance purposes [29–32].

In a 1999 paper, Howland [33] managed to extract Doppler and bearing information from the echoes using the analogue TV video carrier. He further demonstrated the ability to detect and track aircraft at ranges of up to 260 km. In this work, Howland exploited the fact that very accurate Doppler measurements could be obtained when using a stable carrier frequency. As a result, Howland was not concerned with the signal modulation and therefore did not experience the ambiguities encountered by Griffiths in [3].

Howland, like Griffiths, used a dual channel receiver, one channel for the reference and the other for target surveillance. The baseline distance between the Crystal Palace transmitter and the receiver at Pershore was 150 km. He constructed a phase interferometer consisting of a pair of Yagi antennas separated by 0.6λ. As the receiver was situated beyond the LoS of the transmitter, the DSI was substantially reduced. It was also noted that the mutual coupling between the antennas caused inaccuracies in the bearing measurements and had to be compensated for.

Zoeller et al. [34]. found that FM radio signals are attractive illumination signals due to their copious availability, comparatively high transmit powers, and random (noise-like) features, which mean that their AF (depending on suitable programme content) can approach the ideal thumbtack response.

Zoeller found that, due to the low carrier frequency of FM broadcasts, the PR receiver must use long processing intervals to obtain good velocity resolution. The processing intervals experimented with ranged from 0.125 to 0.5 seconds, which equated to velocity resolutions from 3 m/s to 12 m/s [34]. However, processing intervals of 1 to 4 seconds are typically reported in more recent literature [7, 10, 11, 13]. Like most systems, Zoeller utilised a two channel receiver which provided Zoeller with detections of up to 100 km from the receiver. It was however, noted that correct site location for the receiver is critical for the overall system performance.

In 2005, Griffiths and Baker published two papers detailing work analysing the expected performance of PR utilising different transmit waveforms [13, 35]. In these papers, the authors detail the theoretical performance that can be achieved by using FM, DAB, DVB-T, analogue TV and GSM base stations as illuminators, while also evaluating the bistatic radar range equation in the context of PR. In [13], the authors also make mention of the problems related to DSI, providing six different approaches to the suppression of DSI. The authors presented the results shown by Sahr in [36] as an example of how physical shielding can be used to reduce DSI. In [35], the authors demonstrate the usefulness of using the AF as a means to determine the signals suitability for use in PR.

In work reported in 2005, Howland et al. [4] developed a PR that utilised an FM transmitter located at Lopik, approximately 50 km from the receiver. As a result of the shorter baseline compared to the baseline used in [33], Howland suffered a significant increase in DSI compared with that of his previous work. In an attempt to reduce the DSI, a null in the antenna beam was physically steered in the direction of the transmitter.

A significant challenge presented as a result of DSI, is the saturation of the frontend ADC which greatly limits the ability to detect targets. Howland notes that once the DSI was sufficiently reduced and filtered, the data could be processed to search for Doppler and time shifted echoes from potential targets. The receiver bandwidth in

[4] was limited to the effective bandwidth of a single FM channel, resulting in a range resolution of approximately 2 km. While a range resolution of 2 km might seem poor at first glance, the fact that these targets were tracked as far as 150 km from the receiver, illustrates the systems' usefulness [10].

It is well documented that FM transmissions are suited for long range PR coverage due to their large transmit powers and relatively long wavelengths. FM antennas also tend to have wide antenna elevation characteristics when compared to their terrestrial digital counterparts, which allows for better coverage of high altitude targets as demonstrated in [14]. Research at the University of Cape Town (UCT) has focused on FM PR since 2007 [2, 7, 37–45] due to the prevalence of FM radio infrastructure in developing nations, especially across Southern Africa. While FM transmitters remain the most ubiquitous across the globe and offer high powered illuminators of opportunity, the performance of a PR depends heavily on the instantaneous bandwidth of the signal used. The instantaneous bandwidth of FM signals vary depending on their on-air content, which can result in inconsistent performance [46]. This led to the development of a system that combined multiple FM channels in an attempt to improve the overall system performance [47]. In the same paper, Lauri et al. investigated the performance between direct sampling and a super heterodyne receiver architecture, noting that the direct sampling provided performance advantages over the super heterodyne architecture.

Despite having to deal with inconsistent signal bandwidths, FM PR have a number of advantages over often lower power digital signal based PR. These advantages include improved range performance due to high power transmitters in the lower frequency VHF band, high Doppler resolution due to long integration times and possible enhanced Radar Cross Section (RCS) (depending on target shape) at certain geometries due to the relatively long wavelengths of FM signals [11].

2.1.2 General System Limitations

It is clear that one of the biggest technical challenges for continuous wave PR is the presence of DSI, since the target echo can be as much as 90 dB or more below the DSI level [4, 5]. In 2007 Griffiths and Baker published a paper titled "The signal and interference environment in passive bistatic radar" [48]. The paper compared different ambiguity functions of various illuminating waveforms as well as investigating different sources of interference and how to cancel them. They emphasise that up to 80 dB of interference suppression is typically required for reliable target detection. This implies an ADC dynamic range of at least 80 dB or 13 bits is required.

The limitation caused by DSI can be overcome in various ways such as use of physical shielding using separated reference and surveillance channels as was demonstrated by Howland [4], O'Hagan [10], Tong [49], Morabito [50] and Inggs [51]. In a 2012 paper by Tong et al., a Multiple Input Multiple Output (MIMO) based architecture for FM PR is introduced [42]. In [42], the authors demonstrated an equivalent monostatic detection range of 100 km (290 km bistatic) using a transmitter with an ERP of 1.3 kW and employing low cost GPS Disciplined Oscillators (GPSDO) [52, 53] to synchronise the separated receivers.

O'Hagan et al. demonstrated the effect of shielding the surveillance antenna from DSI through some physical means [54] which resulted in a less stringent requirement on the ADC dynamic range. Like Morabito [50], Inggs demonstrated that significant performance gains could be achieved through channelising the signal of interest by using a high sensitivity pre-select filter [55] which was demonstrated to reduce out of band interference by as much as 110 dB. Inggs et al. then demonstrated that a 14 bit channelised narrowband architecture exhibited similar performance to a 16 bit wideband architecture, even though it had 12 dB less theoretical dynamic range [7, 55]. This demonstrates that the difference between a wideband and narrowband frontend receiver architectures can be as much as 12 dB due to the presence of out-of-band interference within the wideband architecture. It must be noted however, that a narrowband architecture such as the one described in [55] limits the number of transmitters that can be used with a single receiver.

Another means of mitigating the effects of DSI is through antenna beamsteering as has been demonstrated by Malanowski in [56] and more recently by Strom in [57]. Tsai et al. [58] investigated the use of an 8 element dipole array to reduce the effects of DSI in an FM PR where a 20 dB null was placed in the direction of the DSI source, resulting in significant performance improvements. In [59] Bournaka et al. investigated the design of a phased array pattern synthesis algorithm for use in DVB-T PR where nulls of up to 40 dB are placed in the direction of either the DSI or other interference sources. Another example of an antenna array being used for null-steering in a DVB-T and DAB PR application is shown by O'Hagan in [60]. Under certain conditions, the authors managed to place a 50 dB null in the desired direction using an 11 element array at 675 MHz, thereby significantly reducing the levels of DSI.

Gould et al. [61] documented the performance and issues relating to a BAE Systems multi-band PR prototype which operated from 100 MHz through to 2 GHz. They developed a 4 channel system which could capture data at 10 MHz, providing enough bandwidth to capture either analogue or digital transmissions in the FM and DVB-T bands. The authors highlight the fact that one of the major difficulties was that of high dynamic range in order to detect targets as low as 90 - 100 dB below the DSI. In order to overcome the issues caused by the DSI, the authors implemented analogue beamforming for null placement and were able to detect targets up to 80 km bistatic range. While beamsteering has been shown to be an effective means of suppressing DSI, most of the work undertaken in the PR context has been with higher frequency digital systems such as DAB, DVB-T and DVB-T2. This is because at the lower VHF band, the size of the antennas for arrays becomes a limiting factor on the null depth.

To supplement physical shielding and null-steering for DSI removal, significant research has been carried out into the removal of DSI in the signal processing domain, with a number of prominent DSI cancellation algorithms being developed. In one of the earlier papers on this type of adaptive filtering, Boray [62] discusses the trade-off between computational complexity and convergence performance in a conjugate gradient based method. This concept was further explored and optimised by Sheng et al. in [63]. A real-time implementation of the Conjugate Gradient Least Squares (CGLS) algorithm was demonstrated by Tong [11] using Commercial Off-The-Shelf (COTS) hardware such as GP-GPUs for use in FM PR. This approach was preferred to another popular algorithm known as Extensive Cancellation Algorithm (ECA) due to the relatively fixed execution time of CGLS [11].

In 2006 Colone [64] described an ECA and its ability to remove DSI from a PR. Colone and O'Hagan et al. expanded this work into what became a widely used and implemented multistage algorithm for disturbance removal and target detection using FM illuminators of opportunity in 2009 [65]. While it is noted that real-time operation was not the objective of the authors, one of the major drawbacks to the ECA algorithm is the long computation times that result from large clutter estimation matrix dimensions, making it impractical to use in real-time systems. Work has been undertaken to improve on the computational efficiency of ECA and in 2016 Chen et al. [66] demonstrated significant computational speed-ups by using a Frequency Modulated Continuous Wave (FMCW)-like batches approach which he called Extensive Cancellation Algorithm - Batches (ECA-B). This batches approach is similar to that demonstrated by Griffiths in [67] and later expanded upon by Petri in [68] that sees a 96% reduction in computational time for the creation of an Amplitude-Range-Doppler (ARD) map. Again, a similar approach was adopted by Ansari in a 2016 paper that proposed the use of a Sequential Cancellation Batch (SCB) algorithm which was shown to have less computational complexity and lower memory requirements than both ECA and ECA-B whilst providing similar cancellation performance [69].

Other, less common DSI cancellation algorithms have been investigated such as Sequential Cancellation Algorithm (SCA), Least Mean Squares (LMS), Recursive Least

Squares (RLS) based algorithms and variations thereof which have been directly compared in the context of PR in [70–74]. Whilst the algorithms mentioned operate in the time domain, highly efficient frequency domain implementations have also been developed as part of the move towards frequency domain based digital Orthogonal Frequency Division Multiplexing (OFDM) waveforms such as DAB, DVB-T and DVB-T2. A frequency domain approach called Extensive Cancellation Algorithm in Carrier and Doppler (ECA-CD) was proposed by Schwark in [75]. In [75] the authors demonstrate the ability to cancel all multi-path clutter, independent of path length, with or without small Doppler shifts by exploiting the signal structure and the range-Doppler processing technique known as inverse filtering. It is important to note that while digital processing has improved significantly, it is not uncommon for both physical shielding and DSI cancellation algorithms to be used in conjunction with each other.

2.1.3   Current Research Focus

Over recent years, there has been significant research into WiFi PR. Early work by Guo et al. in [76] demonstrated the feasibility of using WiFi access points as an illuminator for detecting short range targets in a controlled environment such as an anechoic chamber. Following this work, Chetty et al. [77] published the results of experiments using WiFi as an illuminator in an indoor environment to demonstrate the detection capabilities in high clutter environments. The authors demonstrated the ability to detect human targets above the clutter at walking speeds using highly directional antennas. In 2012 Chetty et al. then demonstrated the ability to detect targets through multi-layered walls [78] which was later expanded upon by Broetjie [79] in 2013 and Wu [80] in 2016 where complete multi-static systems were shown to track human targets for through-wall monitoring and surveillance.

In a paper investigating the use of satellite based DVB-S signals for use as PR illuminators, Sun et al. [81] demonstrated that a theoretical range resolution of 4.95 m could be achieved due to the relatively high bandwidth of 30.27 MHz per channel. In [81] the same authors calculated that with the use of a 0.5 second integration time, a velocity resolution of 0.024 m/s could be achieved, noting that this makes DVB-S highly suitable for 2D moving target detection applications.

DVB-T is, next to FM broadcasts, one of the most common signals of opportunity for passive radar. The second generation of terrestrial Digital Video Broadcast, called DVB-T2, is increasingly being deployed world-wide [82]. Germany, for example, is entirely serviced by DVB-T2. Over the past decade there have been numerous PR demonstrators designed and developed to operate with the original DVB-T broadcast standard [65, 83–90]. Now, however, PRs are being adapted to utilise the increasingly widespread DVB-T2 standard [91–97].

Digital broadcast protocols have the advantage that they have high and constant bandwidth compared to traditional analogue systems. The wider bandwidths provided by digital broadcast services yield finer range resolution, however these systems are not without their drawbacks. DVB-T2 is an evolution of DVB-T and introduces a high level of flexibility in its transmission. This higher level of flexibility enables the standard to be used in a wider range of transmission environments and support for higher data rate transmissions [98, 99].

Like other digital signals, DVB-T and DVB-T2 offer some major advantages for use in passive radar such as high and constant bandwidth as well as the possibility of reconstructing a perfect reference signal using a de- and re-modulation scheme [100–102] commonly referred to as 'demod-remod'. Demod-remod is made possible because the signal follows an open standard, such as [99] in the case of DVB-T2, that can be implemented by anyone. Signal reconstruction, however, is also possible for an adversary, who may exploit the deterministic parts of the signal that are used for synchronization and signalling purposes.

The demod-remod process is discussed by Searle et al. in [103] in the context of DVB-T while O'Hagan et al. demonstrate using demod-remod in the DAB context [100]. In a paper titled "enhancing target detection using real-world data from an (Australian) 8k-mode DVB-T system [89]", Palmer notes that demod-remod process can be used to effectively remove pilot signal ambiguities within an ARD map. Palmer demonstrates that a 36 dB reduction in residual ambiguity peaks is achieved over standard matched filtering with only a 1 dB sacrifice in the zero Doppler, zero delay peak level. In a 2018 paper by O'Hagan and Paine, ambiguity removal in DVB-T2 PR was demonstrated using a demod-remod process [93].

Basic AF analysis of DVB-T2 signals was performed by Pidanic in [104] where DVB-T2 signal with a 4K Fast Fourier Transform (FFT) size is used along with Pilot Pattern (PP) 4. The authors detail the position of the pilot ambiguities however it must be noted that the position of these ambiguities will depend on both the signal FFT size and the pilot pattern in use [93].

In 2010, Baczyk and Malanowski published a paper that demonstrated the process of demod-remod using universal COTS SDR components not dedicated for DVB-T using a chirp-Z transform to estimate the length of the OFDM symbol through autocorrelation [105]. The authors demonstrated the feasibility of this with tests utilising real DVB-T data. While this is a novel approach, it offers little benefit over conventional demod-remod processes. A year later, the same authors expanded on the approach to evaluate its performance from the view of clutter removal and ARD map calculation [88].

Transmitters using OFDM based DVB-T and DVB-T2 signals for PR have been investigated by numerous researchers, most notably [83–86, 89, 90, 102]. Since OFDM signals are digital, they contain periodic structures within the signal itself such as pilot signals. These pilot signals are used for frequency correction and channel estimation and correction which is used in the demodulation of such signals. Unfortunately for PR, these periodic signals cause ambiguities within the ARD map that need to be removed. As a result, there are two widely used approaches to processing OFDM based signals in PR, with the first being mismatched filtering as proposed in [83, 84] and further explored in [86, 89]. Mismatched filtering is a traditional cross-convolution based approach which involves demodulating the reference signal and then remodulating it to obtain a clean, slightly modified reference. The new remodulated reference signal is then used as a 'mismatched' filter when performing the range-Doppler processing.

A second approach adopted by Berger in [85] and expanded by Fang in [90] is referred to as inverse filtering. Inverse filtering is a process whereby the signal undergoes demod-remod to produce a noise free reference signal. This noise free reference signal is then used to perform the range-Doppler processing by first dot dividing the OFDM symbols in the surveillance channel by the same OFDM symbols in the remodulated reference channel. A 2D FFT is then applied to produce an ARD map. This in effect, normalises the carriers and shifts the direct signal clutter into the zero Doppler bin of the ARD map.

A third, less common approach was demonstrated by Polonen et al. in [91] and later again by Cui and Himed in [106] where the authors suggested target detection through the use of the DVB-T2 control symbol. The DVB-T2 signal consists of a P1 symbol which is used for initial synchronisation, course frequency offset correction and P2 parameter descriptions. To achieve this, a unique structure consisting of three distinct parts C, A and B is used. Part A is the main part of the P1 symbol and is 1024 samples long. Part C sits in front of part A to form a cyclic prefix 542 samples long, shifted by one carrier up in frequency. Part B is a cyclic suffix 482 samples long which is also shifted up in frequency by one carrier and appended to the back of part A. Together, the unique C-A-B structure forms the P1 symbol, 2048 samples long which, when correlated with a frequency shifted version of itself, produces a unique impulse response [93, 99]. A further illustration of the signal characteristics is given in Chapter 6. As Polonen demonstrates in [91], the unique P1 symbol can potentially be used to detect the presence of targets without the need for a demod-remod stage. It is however, unlikely that such an approach will work under real world conditions using real world signal levels.

In a 2009 paper by Bongioanni et al., an interesting new approach to ambiguity removal in DVB-T ARD maps was proposed [107]. Rather than removing the ambiguities in the demod-remod stage by normalising the pilots, the authors proposed a computationally efficient approach using an AF based filter that normalises the ambiguities based on their expected positions within the ARD map. This proposed technique is demonstrated to be more robust under certain conditions while being computationally more efficient and does not require strict synchronisation with the reference signal. It is unlikely however, that this approach will be able to adapt to a changing environment such as when there is interference or jamming applied.

Migration from DVB-T to DVB-T2 means that the internal processing of the PR needs to be adapted. In a paper by Winkler et al. [95], the authors highlight that the same basic processing steps used in DVB-T can be deployed for Single Input Single Output (SISO) DVB-T2 networks. However, since the DVB-T2 standard allows for Multiple Input Single Output (MISO) operation, whereby transmitter groups can transmit similar but different content, additional processing is required to remove the interfering transmitter groups. Another characteristic of demod-remod based processing is channel errors. Wojaczek and Cristallini performed a sensitivity analysis regarding the influence of channel errors in mobile DVB-T PR where the effects of inter-channel inequalities within the receiving chain was analysed [108]. The authors noted that while advanced processing techniques such as Displaced Phase Centre Antenna (DPCA) and Space-Time Adaptive Processing (STAP) rely on external calibration, this is not suitable for use in covert PR and as such, proposed digital and post processing based calibration techniques where the direct signal itself is used to calibrate for the inter-channel errors.

2.1.4   Commercial Systems

Over the last two decades, PR have matured to the point where commercial products are being made available. Such systems include Hensoldt's Passive Radar demonstrator (PARADE) [109, 110], the Thales HA-100 [111], the PARASOL system by Fraunhofer FHR [112] and the ComRad3 PR from Peralex Electronics [113]. Older systems such as the Lockheed Martin Silent Sentry that utilised FM radio broadcast transmissions [114] and CELLDAR, developed by BAE Systems and Roke Manor Research [115] to utilise transmissions from GSM mobile phone broadcasts provided an early insight into the commercial viability of PR.

The Fraunhofer FHR PARASOL is the first PR in the world to receive certifying approval from an airspace surveillance and regulatory body, namely the German Deutsche Flugsicherung (DFS) [112]. Passive Radar has also been demonstrated on moving platforms with Warsaw University of Technology (WUT) being one of the first open publications on airborne PR [116, 117]. In 2014, Fraunhofer FHR together with the Australian Defence Science Technology Group (DSTG), then called Defence Science Technology Organisation (DSTO), were among the first researchers to demonstrate the use of PR on a moving maritime platform (a boat) and utilising land-based illuminators of opportunity [118–120]. Presently Fraunhofer FHR is a research leader in the field of Airborne Passive Radar (APR) and in recent years have demonstrated PR SAR and Ground Moving Target Identification (GMTI) [108, 121]. Another innovative and niche PR program was performed by Peralex Electronics in South Africa. Peralex, together with the Square Kilometer Array (SKA) consortium demonstrated the deployment of a PR in proximity of highly-sensitive radio astronomy receivers [94]. The purpose of the deployment was to detect all aircraft in the vicinity of the radio astronomy reserve. Transponder-carrying aircraft pose a particular danger to the sensitive Radio Frequency (RF) circuitry of the radio telescopes. Therefore when an aircraft is detected by a PR, to avoid interference and data corruption, the radio telescope is "informed" (by the PR) so that it can temporarily be de-sensitized, or even shut-down.

The use of multiple DVB-T channels for higher range resolution was demonstrated in a paper by Conti et al. [9]. The authors present two different architectures for exploiting multiple adjacent DVB-T channels, a wideband approach which samples the entire band and a channelised approach which samples each channel individually. In both cases the authors were able to demonstrate that the range resolution performance could be improved by N times with respect to a single channel architecture where N is the number of channels sampled. Regardless of the relative performance of each architecture, it is clear that sampling each channel individually provides better Electronic Protection (EP) as the attacked channel can simply be removed from the ARD map rather than cause significant out of band interference for the open channels.

Modern PRs are increasingly moving away from single illumination sources and towards multi-illuminator operation, often incorporating a blend of FM, DAB, DVB-T or DVB-T2. Schroeder et al. [122] demonstrated such a system in 2012 that utilised FM, DAB and DVB-T with power levels of 50 dBW for FM and DVB-T and 27 dBW for DAB. The system used a 7 element array for the FM and DAB frequency range (88 − 240 MHz) while a 2x7 element array was used for the DVB-T (474 − 850 MHz) frequency range. The system was able to selectively process 8 FM channels simultaneously and the authors noted that targets were detected in both the FM and DVB-T channels up to 38 km bistatic range.

Edrich et al. published a paper in 2014 detailing the design and performance evaluation of a mature FM, DAB and DVB-T PR with the goal of building a system for airspace surveillance with a range resolution comparable to that of an active ATC radar while covering an area with a diameter of 200 km and a target location accuracy of 100 m [123]. The authors mention that the use of FM transmitters allows for greater coverage due to both the high power transmitters $(10 - 100$ kW) as well as the limited down-tilt on the antennas [14]. The DVB-T and DAB transmitters were used as they provided significant improvements in range resolution over the FM PR. The authors were able to demonstrate detections up to and beyond 250 km bistatic range with Automatic Dependent Surveillance - Broadcast (ADS-B) data indicating target altitudes of 10 000 m.

Paine et al. published a paper in 2018 on a multi band FM and DVB-T2 PR that utilises Yagi antennas for the FM band (97.6 MHz) and wire reflector TV antennas for the DVB-T2 band (706 MHz) [94]. The typical FM transmit power in the region was 40 dBW for FM while the DVB-T2 transmit power was 47 dBW. While this is significantly lower than the older 60 dBW analogue TV transmitters used by Griffiths et al. in [3], targets were shown to be detected in both the FM and DVB-T2 channels at bistatic ranges up to 140 km and 36.5 km respectively.

One of the two major challenges remaining for PR in the commercial and military context is the lack of statistical quantification of system performance as is common with active radar systems. Significant research has been undertaken in an attempt to address this and other performance related metrics with O'Hagan publishing his PhD book on the "Performance Characterisation Using FM Radio Illuminators of Opportunity" in 2009 [10]. As recently as 2017, Lysko and Maasdorp published pa-pers describing the efforts of the South African Counsel for Scientific and Industrial Research (CSIR) to fully characterise the FM PR being used for air traffic monitoring in the Johannesburg area of South Africa with the aim of "proving the technology and addressing the needs of ATC in developing countries" [124, 125]. The other major challenge still facing PR is the lack of open knowledge on the performance of a PR in the presence of potential countermeasures.

2.2 Electronic Countermeasures Applied to Passive Radar

This section provides a critique of the limited but important literature relating to ECM in the context of PR. ECM is a widely discussed technology field, however, with regards to research relating to ECM applied to PR, few studies exist in the open literature. Most papers on PR, such as [13] and [126], offer generalisations about the effectiveness of such systems against ECM, and that it can provide excellent ECCM against potential ECM, but almost no evidence is provided in support of such assertions. Part of the problem with ECM in the PR context is the diversity associated with PR, making it a huge task to assess all possible operational scenarios where jamming could be applied.

2.2.1 Passive Radar in the Military Context

North Atlantic Treaty Organization (NATO) Science and Technology Organisation (STO) *Advanced Modelling and Systems Applications for Passive Sensors Group SET-164* produced a comprehensive report on PR performance, focusing particularly on the impacts of clutter [127]. Another NATO group, *SCI-190 Electronic Countermeasures to Radar with High Resolution and Extended Coherent Processing* held a specialist meeting on PR jamming in 2017 that the author of this work contributed towards [127, 128].

A report by Arend G Westra [129] states that "The U.S. military must gain an understanding of passive radar, not merely theoretically, or with minor research and development projects, but with a dedicated effort."

He then notes, "Why build a stealth counter when there is no immediate stealth peer competitor?"

To which he replies, "We cannot afford to spend billions on stealth, only to fail to thoroughly understand and counter rival systems [129]."

It is generally accepted that one of the principal advantages of PR is that it serves as a potential counter to Low Observable (LO) targets. This is due to two aspects of the systems design:

- LO targets are generally designed to present a very low RCS in the monostatic region at microwave frequencies (1 to 10 GHz). Due to the relatively low operating frequency of FM based PR (100 MHz), the longer wavelengths result in an increased RCS of the target.

- The bistatic or multistatic nature of PR results in a situation where the reflected energy is radiated towards the radar rather than away from it as is often the case with monostatic radar and LO targets.

While PR typically exhibits these characteristics, it is unlikely that a PR system will be deployed primarily to counter stealth targets. This is partly due to the difficulty in utilising forward scatter regions for enhanced RCS. In addition to this, if covert operation is not the primary requirement, a multistatic VHF or Ultra High Frequency (UHF) radar will allow for more control and improved system performance.

2.2.2 Electronic Countermeasures Applied to Passive Radar in Open Literature

A rudimentary analysis of the effects of active jamming on an FM band PR is performed by Sendall in a 2016 report [12]. In the report, the author presents an attack where a secondary transmitter is transmitting energy in the same band and therefore unintentionally interfering with the radar system (Page 22). The author models the reference signal as: $r(t) = pk(t) + ql(t) + n(t)$, where $k(t)$ is the normalised signal transmitted by the transmitter of opportunity, $l(t)$ is the normalised signal transmitted by the interfering source (which is assumed to be uncorrelated), $n(t)$ is Additive White Gaussian Noise (AWGN) while p and q are complex weights for scaling of the signals.

The surveillance channel is modelled as: $s(t) = rk(t) + sl(t) + m(t) + n(t)$, where $m(t)$ represents the target echo returns while r and s are complex weights for signal scaling. The author then notes that in order for successful DSI cancellation to be implemented, the ratio: $p/q \approx r/s$, must hold true but since there is likely to be spatial differences between the two sources, the cancellation will be impaired, leading to target masking due to high signal sidelobes. The author concludes that even though the effects of the additional in-band interference, $l(t)$, has not been fully analysed, due to its effect on the reference signal and therefore the cancellation, it can "be established that the presence of an additional, in-band interferer is highly detrimental to system performance." [12].

The most comprehensive, open source of literature on ECM and ECCM applied to PR is provided by Willis et al. in [130] and [131]. In [130], Willis discusses noise jamming and what he calls the 'benchmark range' of the PR. The so called benchmark range of a PR is essentially the root of the product of the transmitter-to-target, receiver-to-target range ($R_B = \sqrt{R_{Tx} \cdot R_{Rx}}$). See equation (6.8) of Chapter 6 of [131] for a full expression.

Once the benchmark range has been calculated, Willis then proceeds to calculate Jammer-to-Noise ratio (JNR) and estimate input noise temperature (using equation (6.54)). Being a noise jammer, the effect on the PR is considered to be an addition to the noise temperature of the receiver. He then divides the benchmark range, R_B, by a factor depending on the ratio of the jammer noise to the system noise to demonstrate the effect on the benchmark range using equation (6.55).

Willis then assumes a specific range difference, $\Delta = R_{Tx} - R_{Rx}$, before finally calculating the resultant receiver-to-target range and coverage area due to the noise jammer using equations (3) and (4) along with Table 6.5 [131] page 129.

To demonstrate, Willis examines an FM PR with a 250 kW transmitter and an effective channel bandwidth of 100 kHz. Two jammers are considered, one with an ERP of 100 W and another with an ERP of 1 kW. The results show that the PR is very vulnerable to jamming in the mainlobe of the receiver but considerably less so when the jamming is applied to the sidelobe of the receive antenna. This further emphasises the idea of spatially diverse receivers as a counter to ECM. In the same book Willis also dispels the idea that the jammer can be used as an illuminator of opportunity for the PR by illustrating that the power levels are simply too low [131].

In a paper by Zheng [71] the authors discuss accidental jamming of an FM PR. They point out that DSI cancellation techniques have been discussed by many authors, with most implemented algorithms adopting a single-stage adaptive filter processor structure to remove or cancel the DSI. It was shown that in practice, because of the complicated atmospheric propagation influence (such as maritime evaporation ducts), or even active radio jamming, a single-stage DSI canceller could potentially be invalid.

Zheng noted that in practice, due to atmospheric propagation effects and/or active radio jamming, a two-stage canceller should be used to combat the DSI and jamming effects on the system. They report on field experiments and simulations to validate the approach, demonstrating a jamming suppression of more than 20 dB, achieving desired performance through cancellation of the DSI signals from the selected transmitter.

In one of the few journal papers on the subject, which builds off the work in [45], Inggs et al. [2], demonstrated the ability of a noise jammer to effectively jam an FM PR provided the receiver location is known. Through the simulation package Flexible Extensible Radar Simulator (FERS) [132, 133], Inggs demonstrated that a simple 10 W noise jammer could be used to completely mask a potential target, however, it is clear that this approach is simplistic in that without looking at the specific JSR for a given scenario, the jammer ERP is largely meaningless. Inggs also pointed out that the work performed by Willis in [131] is an oversimplification since only looking at the change in Signal-to-Noise ratio (SNR) would not account for the highly non-linear effects of the DSI cancellation algorithms and Constant False Alarm Rate (CFAR) filters. It was however mentioned, that systems utilising multiple distributed receivers would be significantly more difficult to counter.

It is further concluded that in [2]:

- If the Rx sites are known, the PR is very vulnerable to a jammer with very modest power levels (1 to 10 W).

- Cancellation techniques seem unlikely, since if they work at high fidelity, they will also remove the signal.

- If the receiver sites are unknown, it will be very difficult for the ECM operator to ensure that sufficient jammer power is provided to all the receiver sites.

- It seems that any signal that fills the PR bandwidth will be successful in jamming.

- Spatially diverse receivers will be very difficult to jam.

- Simple null steering is an effective countermeasure to stand-off jamming.

- Self protection jamming will be very difficult to mitigate.

- Considerable research into various aspects of each jamming technique still needs to be performed.

Unlike analogue based FM PR that have no strictly deterministic components, digital OFDM based PR such as DAB, DVB-T and DVB-T2 have deterministic components that can be exploited by a potential attacker. This was demonstrated by Schüpbach et al. [134] when the Phase Reference Signal (PRS) was exploited to successfully attack a DAB PR. It was noted that due to the robustness of the DAB signal, attacking the reference channel to prevent reconstruction would require jammer power of the same order of magnitude as the transmitter of opportunity, making this an undesirable form of attack. The authors demonstrated the effectiveness of utilising coherent jamming over noise jamming on the surveillance channel for these types of systems due to significant integration gain that can be achieved.

This approach was taken further by Schüpbach and Paine et al. in [135] where it was shown that the deterministic components of the DVB-T signal, namely the pilot carriers, could be exploited to effectively attack the PR with relatively low power levels.

This demonstrates that the deterministic components within the reference signal can be exploited and used as attack vectors in OFDM based PR.

In a paper by Giusti et al. [136], the authors demonstrate self-protection jamming through the use of a target-on-board DRFM based attack to insert false targets into an OFDM based imaging PR. The authors begin the paper with the broad statement that PR is robust against jamming because the receiver location is unknown. It is noted that since PR systems utilise broadcast transmitters, operating in frequency bands requiring a licence that forbids the transmission of interfering signals, conventional jammers cannot be used. This is a largely moot point since in a hostile environment, the idea of adhering to spectrum licensing is largely ignored.

The paper then states that the use of OFDM based signals as a reference signal further improves robustness against jamming due to the random nature of the OFDM signal. This is incorrect and is in direct contrast to what has been demonstrated by Schüpbach and Paine et al. in [134] and [135] where it was shown that the deterministic nature of the OFDM signal can be used to achieve additional processing gain to improve jamming performance. In [134] and [135] the authors demonstrated the ease at which an OFDM based DAB or DVB-T PR could be jammed or spoofed using a self protection DRFM based attack.

While DRFM based attacks are common in more advanced ECM systems used against traditional active radar, little work has been done on their application to PR. Continuing with the assumption from [136] that the DRFM signal should not interfere with the broadcast transmitters, one of the advantages of using a DRFM based attack is that the low power DRFM signal cannot easily be intercepted since it is hidden by the high powered transmitter signal.

As was noted in [134] and [135], the authors of [136] point to the fact that while the DRFM signal delay can be controlled to appear near the true target echo, the Doppler shift cannot be due to the unknown location of the receiver. This can, however, be overcome by transmitting multiple false targets at various different Doppler shifts in an attempt to overwhelm the receiver. This is demonstrated through the use of simulations whereby false targets are placed in the vicinity of the true target echo, resulting in reduced detection and classification performance of the imaging PR.

It is clear from the few available open literature that PR are not as robust against ECM as originally thought, provided the receiver location is known to the attacker. However, as PR receivers emit no EM energy, one of the greatest difficulties in jamming and deceiving PR is that the receiver location is unknown and can be difficult to locate. A technique to locate PR receivers was proposed by Hoyuela et al. [137]:

- Satellite imaging radar (eg. SAR).

- Electro-optic sensors.

- 'Realistic ideal' location estimation.

Therefore, to avoid using kilowatts of jamming power, an additional means of intelligence has to be used. For the detection of PR receivers, surveillance with imaging radar or electro-optic cameras to detect the relatively large antennas is one possibility. Research has been undertaken on the optimisation of the receiver location for given performance requirements. Often the optimised result is unrealistic so Digital Terrain Elevation Data (DTED) and Geographic Information System (GIS) data can be utilised to establish the 'realistic ideal' receiver location [42, 44, 137–139].

Tactics employed in active radar such as relocation and transmit waveform agility are not feasible with PR since the receiver has no control over the transmitter and the transmitter infrastructure can be easily and cheaply destroyed. To combat this, a dual mode Active Fall-back Component (AFC) was proposed by O'Hagan et al. in [140] whereby if anything were to happen to the transmitter infrastructure, the AFC could be used as a supplement. Ideally, in a conflict environment, the PR sensor would remain as silent as possible to remain hidden to reduce the risk of it being targeted while utilising as many available transmitters as possible to maximise robustness against transmitter attacks. However, all transmitters are vulnerable to some form of attack, be it physical or electrical whether the receiver is co-located or not. Examples of such attacks include:

- On April 23rd, 1999, the Serbian state television headquarters in Belgrade was destroyed by NATO bombing.

- Between March 24th, 1999 and June 10th, 1999, bombings reduced the Serbian broadcast infrastructure.

- Broadcast infrastructure also destroyed in Iraq, Afghanistan, Libya, Syria and other places.

- Broadcast infrastructure can be destroyed in many ways (asymmetric vulnerability), e.g. Lightning, Sabotage, Terrorism, War, Cyber attack.

- Elimination of power grid.

This emphasises the need for a PR to utilise multiple transmitters and multiple site locations in order to at the very least ensure redundancy.

2.3 Chapter Summary

To conclude this literature review, there is clearly a gap in the literature with regards to ECM applied to PR. As noted, there is little to no open literature available on ECM against PR, with the only relatively comprehensive open research being performed by the UCT Radar Remote Sensing Group (RRSG), found in [2]. This chapter has discussed various technical aspects regarding the use of PR. Also covered is a brief overview of past and present PR while highlighting PR ECM specific literature.

Regardless of the implementation of future ATM systems, such as ADS-B or multilateration, there will still be a requirement for Primary Surveillance Radar (PSR) coverage by primary radar due to national defence and system security needs. PSR will be used as a back-up system and as a means of tracking non-cooperative users. A Civil Aviation Authority (CAA) roadmap for implementation of changes to surveillance over the next 20 years plans the "introduction of Multi-Static PSR to replace primary radar" in 2015 to 2020 and beyond 2020, the CAA plan a "Wider roll out of Multi-Static PSR." [19].

With sensors becoming more and more prevalent in the military context and civilian context, this **book** aims to fill the important gap in open literature regarding ECM and ECCM against PR. All of the limited available literature is based off simulated results with no real world data to verify these claims. As a result, this **book** focuses specifically on ECM applied to FM and DVB-T2 based PR and provides real world representative measurements to validate the simulated results.

Chapter 3

FM Passive Radar

This chapter begins with an overview of the FM radio signal structure and then leads to an investigation of the performance of a single node FM radio based PR in the presence of jamming and quantifies the effect of DSI cancellation. A complete jammer waveform study is presented and the most effective jamming waveform is investigated. A representative measurement from a real FM radio PR is presented as a validation of the simulated results.

3.1 FM Signal Overview

The mathematical description of broadcast FM radio signals is developed in this section. The mathematical description follows that by Stremler in [141]. Frequency modulation is a form of analogue modulation where the baseband information carrying signal, typically called the message signal, $m(t)$, modulates the frequency of the carrier wave. Broadcast FM radio signals are generated by applying a message signal to a Voltage Controlled Oscillator (VCO). The output of the VCO is a constant amplitude sinusoidal carrier wave whose frequency is a function of the control voltage, $m(t)$. When no message signal exists, the carrier wave is simply at its centre frequency, f_c. When a message signal exists, the instantaneous output signal varies about the carrier frequency as expressed by:

$$f_i(t) = f_c + K_{VCO} \times m(t) \tag{3.1}$$

where K_{VCO} is the voltage-to-frequency gain of the VCO, expressed in units of Hz/V. The resultant output of $K_{VCO} \times m(t)$ is the instantaneous frequency deviation, Δf. The instantaneous phase of the signal is equal to 2π multiplied by the integral of the instantaneous frequency, giving:

$$\theta_i(t) = \int_0^t f_i(t)dt$$
$$= \int_0^t 2\pi f_c dt + \int_0^t 2\pi K_{VCO} \times m(t)dt \tag{3.2}$$
$$= 2\pi f_c t + 2\pi K_{VCO} \int_0^t m(t)dt$$

The output FM waveform, $X_{FM}(t)$, is therefore represented by:

$$X_{FM}(t) = A_c \cos(\theta_i(t)) \tag{3.3}$$

If the initial phase is assumed to be zero for simplicity, the output FM signal becomes:

$$X_{FM}(t) = A_c \cos\left[2\pi f_c t + 2\pi K_{VCO} \int_0^t m(t)dt\right] \tag{3.4}$$

where the frequency modulated output, $X_{FM}(t)$, has a non-linear dependence on the message signal, $m(t)$, making it difficult to analyse the exact properties of an FM signal. The baseband message signal, $m(t)$, can be modelled mathematically as a sum of each channel component:

$$
\begin{aligned}
m(t) = {} & C_0[L(t) + R(t)] \\
& + C_1 cos(2\pi \times 19 \text{ kHz} \times t) \\
& + C_0[L(t) - R(t)]cos(2\pi \times 38 \text{ kHz} \times t) \\
& + C_2 RDS(t)cos(2\pi \times 57 \text{ kHz} \times t)
\end{aligned}
\tag{3.5}
$$

where C_0, C_1 and C_2 are the gains used to scale the amplitudes of the left channel audio, the right channel audio, the 19 kHz pilot tone, and the Radio Data System (RDS) subcarrier, respectively, to generate the appropriate modulation index, β. For simplicity, the FM signal bandwidth can be estimated by representing the message signal, $m(t)$, by a single tone:

$$m(t) = A_m cos(2\pi f_m t) \tag{3.6}$$

where A_m is the amplitude of the message signal and f_m is the message tone frequency.

Substituting (3.6) into (3.4) gives:

$$
\begin{aligned}
X_{FM}(t) &= A_c \cos\left(2\pi f_c t + \frac{K_{VCO} A_m}{f_m} \sin(2\pi f_m t)\right) \\
&= A_c \cos\left(2\pi f_c t + \frac{\Delta f}{f_m} \sin(2\pi f_m t)\right) \\
&= A_c \cos(2\pi f_c t + \beta \sin(2\pi f_m t))
\end{aligned}
\tag{3.7}
$$

The peak frequency deviation, $\Delta f = K_{VCO} A_m$, is a result of the message amplitude and gain of the VCO. The ratio of the peak frequency deviation, Δf, and the message signal frequency, f_m is known as the modulation index, β. The number of significant sidebands in the output spectrum is a function of the modulation index. This can be seen by representing the FM output signal in terms of nth order Bessel functions of the first kind:

$$
X_{FM}(t) = A_c \sum_{n=-\infty}^{\infty} J_n(\beta) \cos(2\pi(f_c + nf_m)t)
\tag{3.8}
$$

By taking the Fourier transform of (3.8), the discrete FM spectrum can be represented in magnitude by coefficients as a function of β:

$$
X_{FM}(f) = \frac{A_c}{2} \sum_{n=-\infty}^{\infty} J_n(\beta) \left[\delta(f - f_c - nf_m) + \delta(f + f_c + nf_m) \right]
\tag{3.9}
$$

The number of sidebands of an FM signal and its associated magnitude coefficient can be found with the help of Bessel function tables [142, 143].

The average power envelope of an FM signal is constant and can be determined using the Bessel functions as:

$$
P_{ave} = \frac{1}{2} A_c^2 \sum_{n=-\infty}^{\infty} J_n^2(\beta)
\tag{3.10}
$$

As the bandwidth of the message signal reduces, the number of significant sidebands required for transmission reduce. This leads to an increased level of the carrier component, J_0. As the bandwidth of the message signal increases however, the number of significant sidebands required for transmission increases, i.e. J_n for $n > 0$ increases, and the level of the carrier tone, J_0, subsequently decreases.

Figure 3.1 illustrates the FM radio station spectrum output for high, medium and low bandwidth message signals. Each signal is normalised to the maximum level within each plot. From (3.10), the average power envelope across the channel must remain constant. It is therefore clear by comparing the high bandwidth signal (left) to the low bandwidth signal (right), that the carrier tone levels are significantly higher for a lower valued β than for high values. In the context of using FM radio signals as illuminators of opportunity, it is important to note that the integration gain achieved through matched filtering is directly proportional to the instantaneous signal bandwidth. If the message bandwidth occupied the maximum permissible FM radio station bandwidth [2], the corresponding maximum integration gain would be 59 dB (assuming for the time being an integration time of 4 s), as calculated in (3.11).

$$
\begin{aligned}
G_{int} &= t_{int} \cdot \text{B} \\
&= 10\log(4 \text{ s} \cdot 200 \text{ kHz}) \\
&= 59.03 \text{ dB}
\end{aligned}
\tag{3.11}
$$

A 59 dB integration gain can only be achieved when the entire modulation bandwidth is filled for the duration of the Coherent Processing Interval (CPI). Realistically, this will only be the case when applying broadband noise jamming to the PR receiver. A more realistic integration gain for the FM based PR receiver under normal operation is determined by the average modulation bandwidth across the CPI. This results in an integration gain of approximately 55 dB as shown in (3.12).

$$
\begin{aligned}
G_{int} &= t_{int} \cdot \text{B} \\
&= 10\log(4 \text{ s} \cdot 80 \text{ kHz}) \\
&= 55.05 \text{ dB}
\end{aligned}
\tag{3.12}
$$

$$
\therefore 55.05 \text{ dB} < \text{G}_{int} < 59.03 \text{ dB}
\tag{3.13}
$$

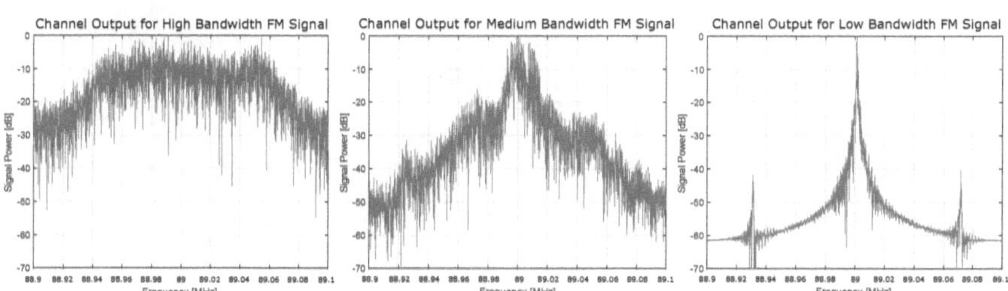

Figure 3.1: FM channel output for high, $\beta = 4$ (left), medium $\beta = 2$ (middle) and low $\beta = 0.25$ (right) bandwidth message signals.

3.2 Typical FM Passive Radar Processing Chain

The exact processing chain used in FM PRs varies depending on system architecture. However, the underlying process is the same for most architectures. An overview of the processing chain used for this investigation is presented while a more comprehensive discussion on FM PR processing chains can be found in [11]. Figure 3.2 illustrates a block diagram of the processing stages involved.

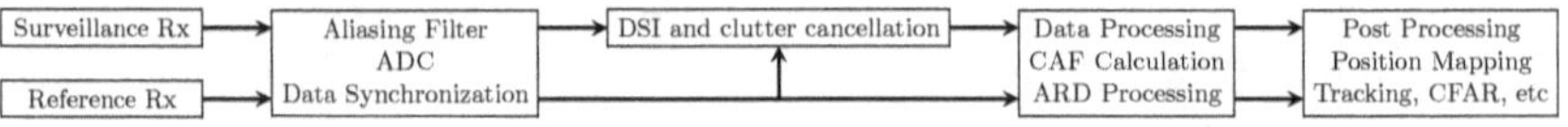

Figure 3.2: Basic passive radar processing chain [11].

As is common with most FM PR, there are two receive channels, one for reference and one for surveillance. The reference channel is directed towards the transmitter of opportunity and is used to record the illuminating signal. The surveillance channel is directed away from the illuminating transmitter and towards the desired coverage area. The raw data is sampled and filtered, usually using a frontend pre-select filter stage to remove out-of-band interference. It is important that the two channels are synchronised to allow for coherent processing.

As discussed in Chapter 2, DSI cancellation is critical for FM PR performance. While null-steering of the antenna helps to improve channel separation that results in a reduction in the DSI in the surveillance channel, an additional software based DSI cancellation process is typically also used. As mentioned in Chapter 2, there are many different DSI cancellation algorithms available with most resembling some form of LMS based filter. This work demonstrates the use of two different DSI cancellation algorithms namely: ECA and CGLS, the exact details of which are described in Appendix B.

Once the signal is recorded, the DSI is removed from the surveillance channel and range-Doppler processing is performed to produce an ARD map. Range-Doppler processing in FM PR is achieved using a typical Cross multiply - Fourier transform (XF) or Fourier transform - Cross multiply (FX) based process. Due to there being more Doppler bins than range bins in the output ARD maps, it is computationally more efficient to use the XF process, however, the resultant output is the same regardless. The output ARD map is then passed to a Greater of Cell Averaging (GOCA) CFAR filter that extracts target detections above a pre-defined threshold. The theory of CFAR filters has been extensively documented in literature [144–148] and will not be discussed in detail here.

The exact parameters of the CFAR filter change depending on the system environment but typically 4 guard and 8 reference cells are used around the Cell Under Test (CUT) with a desired probability of false alarm, P_{fa}, of 10^{-5} in FM PR systems. A GOCA CFAR filter is chosen as it has been proven reliably reject clutter ridges which are common due to the mountainous topography surrounding the Western Cape of South Africa [11]. The CFAR filter is applied in the Doppler dimension due to the large bandwidth and therefore range fluctuations from CPI to CPI in an FM PR. While not implemented for the purpose of this investigation, the next step is for the CFAR detections to be handed to a tracking filter.

3.3 Jammer-to-Signal Ratio

Simulations are described in Section 3.5 with the aim of quantifying the performance of an FM PR in the presence of various jamming signals. However, before these simulations can be performed, appropriate performance metrics need to be determined. One of the challenges to quantifying the performance of a PR in the presence of jamming is attempting to quantify the PR itself. The required jammer power depends on the relative positions of the target, illuminator of opportunity and all the relevant link budgets, making it difficult to quantify jamming performance in a generic way. As a result, the JSR at each Rx channel, prior to integration gain, is used as a performance metric. the JSR is determined using the one way range equation for the jammer power (3.14) and the two way bistatic range equation for the signal echo power (3.15).

$$J = \frac{P_j G_j G_r \lambda^2}{(4\pi R_{jrx})^2} \tag{3.14}$$

$$S = \frac{P_t G_t G_r \lambda^2 \sigma}{(4\pi)^3 R_{tx}^2 R_{rx}^2} \tag{3.15}$$

where,

$$
\begin{aligned}
J &= \quad \text{Direct path jammer signal at surveillance receiver.} \\
P_j &= \quad \text{Jammer transmit power.} \\
G_j &= \quad \text{Jammer antenna gain.} \\
G_r &= \quad \text{Receive antenna gain.} \\
R_{jrx} &= \quad \text{Distance from jammer to receiver.} \\
\lambda &= \quad \text{Free space wavelength } (c/f_c) \\
S &= \quad \text{Target echo power at the surveillance receiver.} \\
P_t &= \quad \text{Transmitter of opportunity transmit power.} \\
G_t &= \quad \text{Transmit antenna gain.} \\
R_{tx} &= \quad \text{Transmitter to target distance.} \\
R_{rx} &= \quad \text{Target to receiver distance.} \\
\sigma &= \quad \text{Bistatic RCS of potential target.}
\end{aligned}
$$

Resulting in a Jamming-to-Target-Echo ratio (JSR_E) on the surveillance channel of,

$$\frac{J}{S} = \frac{P_j G_j G_r \lambda^2 (4\pi)^3 R_{tx}^2 R_{rx}^2}{P_t G_t G_r \lambda^2 \sigma (4\pi R_{jrx})^2} = \frac{P_j G_j 4\pi R_{rx}^2 R_{tx}^2}{P_t G_t \sigma R_{jrx}^2} \tag{3.16}$$

When calculating the JSR for the reference channel, the equation is modified slightly. The one way propagation path loss from the transmitter to the receiver using (3.14) is compared to the one way propagation path loss from the jammer to the receiver*. Once an appropriate representation for the JSR has been determined, an appropriate JSR is chosen to compare the performance across all simulations as shown in Section 3.5 below. The jammer power before antenna gain in each simulation was therefore chosen to be 5 W. This was chosen as it provides a practical power level for a real world jammer. Increasing the power level of the jammer would allow the jammer to be placed at further distances from the receiver while maintaining the same JSR.

$$JSR_{surveillance} = 58.9 \text{ dB}_{\text{start}} \text{ to } 52.6 \text{ dB}_{\text{end}} \tag{3.17}$$

$$JSR_{reference} = -31.6 \text{ dB} \tag{3.18}$$

Equations (3.17) and (3.18) show the JSR at each channel across the simulations. The JSR on the surveillance channel is the ratio of jammer power to target echo power and therefore it is dependent on the target position within the simulation. At the start of the simulation, when the target is furthest from the receiver, the JSR is its highest at 58.9 dB while at the end of the simulation when the target is at the closest point to the receiver, the JSR is at its lowest at 52.6 dB. The JSR on the reference channel remains at -31.6 dB however, as it is determined by the ratio of the jammer power to the reference transmitter power and is therefore independent of the target position.

3.4 Discussion of CFAR in the Context of PR

Unlike false alarms in traditional active monostatic radar, false alarms in PR, and particularly in FM PR, are not statistically well defined. In an active system, a threshold above the noise (in a noise limited system) can be defined from the Receiver Operating Curves (ROC) corresponding to a desired P_d and P_{fa} for a given SNR.

This is not the case for a PR system since the receiver has no control over the transmit-

---

*The assumption is made that the gain of the reference antenna in the direction of the jammer is 0 dBi as the jammer is pointing into a side lobe of the reference antenna and the main lobe of the surveillance antenna.

ter or the transmitted waveform. In the case of an FM PR, the transmitted waveform is continuously changing in a random fashion, resulting in variable system performance. Digital PR systems such as DVB-T2 provide a more consistent performance measure and while work has been done to quantify the performance of PR systems [10], significant work is still required to statistically characterise each system in a given environment.

In an active radar system, the receiver has complete control over the transmitter and the transmitted waveform. The transmitted waveform can therefore be dynamically adjusted along with the CFAR parameters in the presence of interference. This would be done to maintain the desired P_d and P_{fa} in the presence of a dynamic interference environment. With PR, there is no such control of the transmitter or the transmit waveform and so the CFAR is established locally and is not dynamic.

Furthermore, the zero-Doppler direct signal introduces additional artifacts and while it is largely suppressed, it may both directly or indirectly mask targets at lower Doppler. The very act of suppressing DSI can reduce the target SNR.

3.5   System Simulation Parameters

As shown by Inggs et al. in [2] and [55], AWGN can be an effective means of jamming an FM PR. It was also mentioned by Sendall in [12] that any potential jamming would have an adverse effect on the DSI cancellation and therefore the detection performance of the system. To investigate these claims and to further examine the effect of the DSI canceller when jamming is applied, numerous simulations were performed and the results analysed.

The FM PR was simulated using FERS [132], a powerful radar simulator developed at UCT that allows for an arbitrary number of transmitters and receivers. For the simulations shown here, a single FM band (88 − 108 MHz) transmitter-receiver pair was modelled while two system configurations were investigated, the first being a co-located system (whereby both the reference and surveillance antennas were at the same location) and the second being a separated system (whereby the reference and surveillance antennas were significantly spatially separated).

To allow for consistency across simulations, the system specifications were kept constant across each run-cycle. Table 3.1 outlines the simulated system parameters used in this investigation. Figure 3.3 illustrates the simulated system geometry at the start of the simulation. Figure 3.4 depicts a plan view of the simulated system geometry corresponding to the actual system geometry used. The scene depicts the Western

Cape of South Africa. The simulated parameters mimicked real world placement of each component such as height and location in 3D space. It must be noted however, that due to limitations in the FERS simulation package, no clutter or terrain mapping is simulated and therefore DSI is only present for the first few range bins.

Each simulation uses real recorded FM signals for the transmitter of opportunity. The FM radio signal was recorded at UCT and therefore provided an accurate signal model. Each simulation is a 3 minute scenario, allowing the target to descend from 10 000 m to 5 000 m as it approached the Cape Town International airport (CPT) (represented by the two yellow pins in Figure 3.4). The jammer was placed on the top of Tygerberg hill to allow for optimal coverage and its main beam aimed towards the airport to mimic a possible tactical deployment. Unless otherwise stated, each simulation output was processed using the CGLS method of removing DSI as this was the technique that was implemented in the practical system, using real world parameters, available at UCT. CGLS has inherent benefits in that real time processing can be achieved over long processing intervals while maintaining a fixed memory footprint within the processing chain [11, 41].

To maintain consistency with each of the results, the ARD map that was shown for each simulation was from the same, one minute interval into the simulated flight. Unless otherwise stated, each ARD map is normalised to zero where zero represents the maximum peak within the plot. Figure 3.5 shows a two second spectrum output of the signals used in the FERS simulations. The CPI for each simulation was set to 4 seconds as this was in line with what is used in the real systems [11].

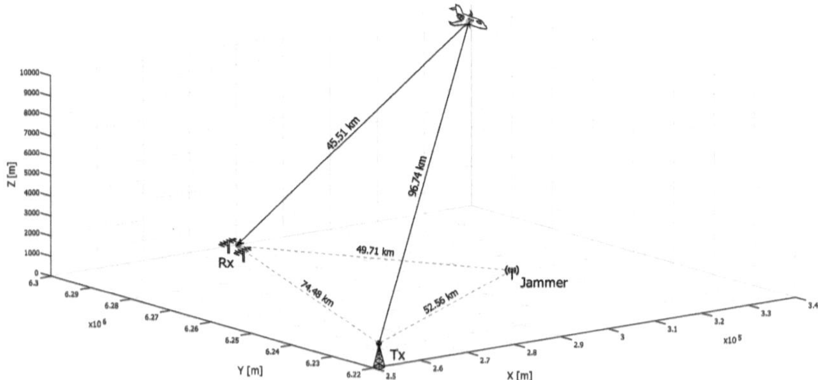

Figure 3.3: Cartesian geometry entered into FERS for the start of the simulated flight path.

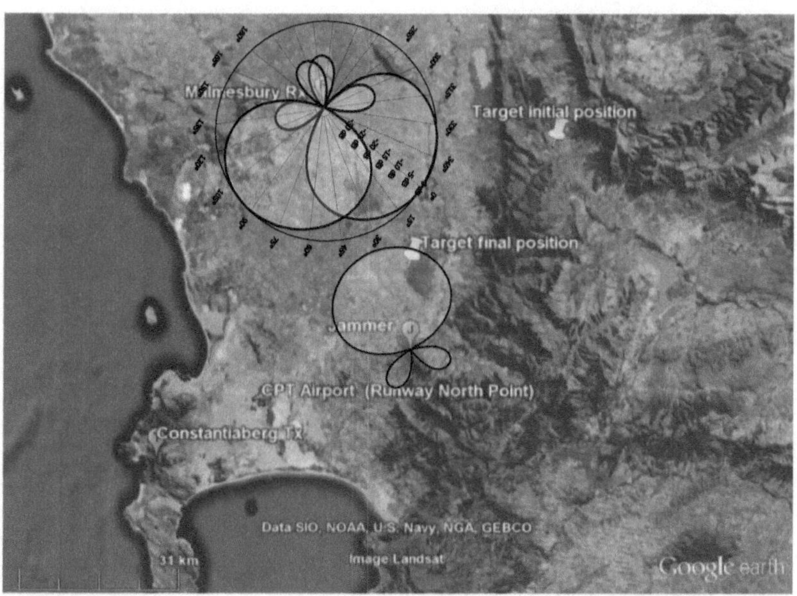

Figure 3.4: Overhead view of scenario geometry used in each FERS simulation with antenna beam patterns overlayed [55].

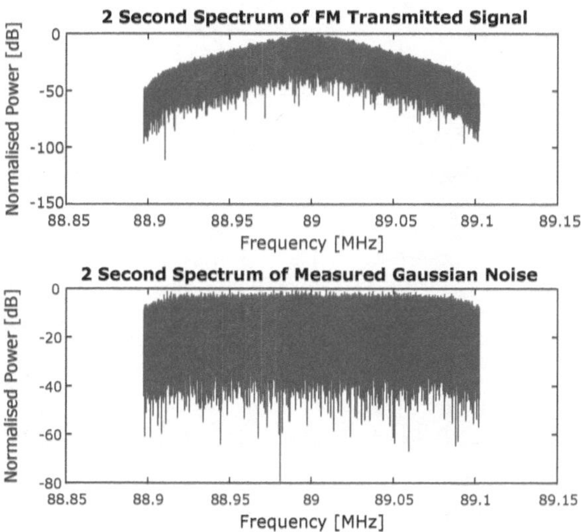

Figure 3.5: (Top) Two second spectrum of the real recorded FM signal that is used as the transmit signal in the FERS simulations. (Bottom) Two second spectrum of the in-band real recorded Gaussian noise signal that is used as the jamming signal in the noise jamming FERS simulations.

Table 3.1: Simulation parameters used in each of the FERS simulations.

Transmitter (Tx)	
Antenna Beam Pattern	Isotropic
Antenna Gain	0 dBi
Antenna Altitude	400 m
Carrier Frequency	89 MHz
ERP	10 kW
Waveform	Real recorded FM data, 204.8 kSps complex sampled
Reference and Surveillance Receivers (Rx)	
Antenna Beam Pattern	Sinc
Antenna Gain	7.2 dBi
Antenna Altitude	240 m
LO Error	50 ppb (std. dev. of 0.01 Hz @ 204.8 kSps)
Noise Figure	4 dB
Digitisation	204.8 kSps complex, 16 bit quantisation
Target	
Initial Altitude	10 000 m
Final Altitude	5 000 m
Velocity	Constant 200 m/s
RCS @ 89 MHz	23 dBsqm (200 m^2 a large airliner)
Swerling Case	0 (Non-fluctuating)
Jammer	
Antenna Beam Pattern	Sinc
Antenna Gain	7.2 dBi
Transmit Power	5 W before antenna gain
Carrier Frequency	89 MHz
Waveform	AWGN, Sine wave on carrier
Processing Parameters	
DSI Cancellation	5 range, 5 Doppler bins
DSI Cancellation CPI	102400 samples (0.5s)
Range/Doppler Processing	120 range, 1601 Doppler bins
Range/Doppler CPI	819200 samples (4s)
CFAR Algorithm	GOCA-CFAR
CFAR Window	4 guard cells, 8 reference cells (either side of CUT)
CFAR Dimension	Doppler (Robust against bandwidth fluctuations)
CFAR Threshold	$P_{fa} = 10^{-5}$ (exponential noise model)

3.6 Quantifying the Effect of a DSI Canceller

This section investigates the effects of a DSI canceller by first injecting AWGN into
both reference and surveillance channels as would be the case with a co-located system
and then into only the reference or only the surveillance as would be the case with a
separated reference and surveillance system.

Using the simulation parameters described in Section 3.5, a clean simulation, i.e. one that contains no jamming, is run to provide a performance benchmark for the system in absence of jamming or other unwanted interference, the results of which are demonstrated in Figure 3.6.

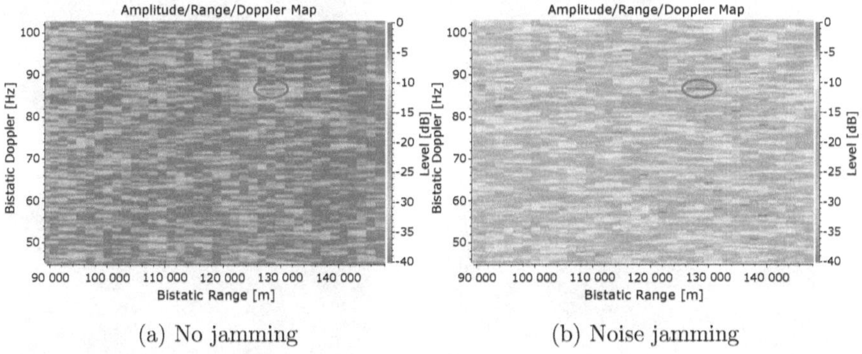

(a) No jamming   (b) Noise jamming

Figure 3.6: FERS simulated target one minute into the 3 minute long simulated flight path with jamming applied to both channels. Target location is located in the red oval. The increased noise floor as a result of the jamming can be seen by the 22 dB reduction in dynamic range between the target and the surrounding noise floor.

Once it was clear that the target was detected when no jamming was present, simulations were run to determine the effects of applying broadband noise jamming in each channel of the PR. The first jamming simulation is completed by applying the jammer to both the reference and surveillance channels, as would be the case in the scenario demonstrated in Figures 3.4 and 3.3 with co-located channels. Figure 3.6b illustrates the result of applying 5 W of noise jamming across the FM channel where the same target, clearly visible in Figure 3.6a, was masked by the raised noise floor.

Passing the output of the ARD maps into the CFAR filter described in Table 3.1 results in an accumulated output as shown in the top left corner of Figure 3.7. It illustrates the resultant combined output of the CFAR filter for the simulated system with jamming applied to the different channels. The results demonstrated in Figure 3.7 indicate that jamming performance heavily depends on two factors, namely: which channel (reference or surveillance) of the PR is affected by the jamming and the performance of the DSI canceller to suppress the jamming signal.

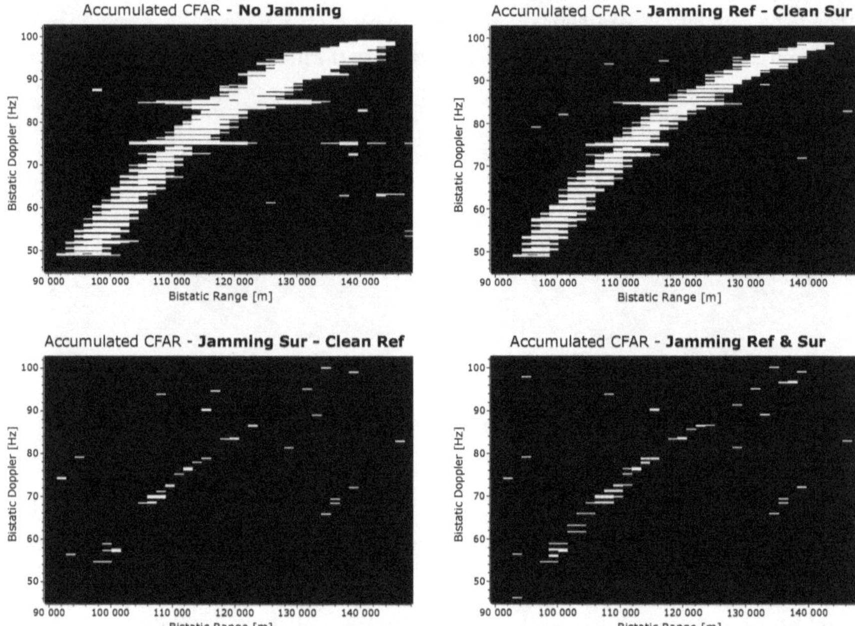

Figure 3.7: Combined CFAR output over the entire simulated flight showing the comparative system performance in the presence of jamming. (Top Left) Clean simulation, no jamming applied. (Top Right) Jamming applied in only the reference channel and not in the surveillance channel. (Bottom Left) Jamming applied in only the surveillance channel and not in the reference channel. (Bottom Right) Jamming applied in both the reference and surveillance channels.

The performance of the system in the presence of jamming is defined by the number of target detections, N_d, achieved while jamming is applied, compared to the number of target detections achieved when no jamming is applied (where $N_d = 100\%$). Observing Figure 3.7, the top left plot shows the accumulated CFAR output for the target over the full 3 minute flight simulation. This results in 45 ARD maps (180 seconds divided into 45×4 second plots) being filtered and a target being detected in each plot. Applying a jamming signal to the reference channel, typical for systems that utilise separated reference and surveillance channels [42, 49], shows no visible reduction in the number of target detections at the output of the CFAR filter. This is as a result of the jammer signal being 31.6 dB lower than the reference signal. Applying the same jamming signal to the surveillance channel, while maintaining a clean reference channel results in a drastic reduction in system performance with N_d dropping to 29%. Injecting the same jamming signal into both the reference and surveillance channels, as would be typical for a system that utilises co-located antennas (such as the one described in

Figures 3.4 and 3.3), results in a relative improvement in detection performance such that $N_d = 44\%$ compared to $N_d = 29\%$ when jamming only the surveillance channel.

It is clear from the results demonstrated in Figure 3.7 that applying the jamming signal to only the reference channel is the least effective, while applying the signal to only the surveillance channel is the most effective. Applying a jamming signal to both the reference and surveillance channels sees a reduction in jamming performance when compared to jamming of only the surveillance channel. When the jamming signal appears in both the reference and surveillance channels, as would be the case with a co-located system, the target is found to be detected more consistently than with only the surveillance channel being jammed. This result is attributed to the DSI cancellation step in the processing chain. If the jamming signal is found in both the reference and surveillance channels, the DSI canceller, which suppresses the reference signal in the surveillance channel, suppresses the added jamming signal as it appears as part of the reference (since it appears in both channels). The canceller then attempts to remove the jamming signal and in doing so, slightly reduces its effect on the system. The reason that the jamming signal is not fully removed is due to the fact that it is at such a low level when compared to the reference signal itself, as shown in (3.18). For the DSI canceller to remove significantly more of the jamming signal, it would require a very sharp and deep null, leading to increased computation time and system complexity.

As discussed in Chapter 2, Zheng et al. [71] describe how they were able to achieve more than 20 dB of interference cancellation using a two stage canceller to combat inadvertent jamming. It is clear from Figure 3.7 that the DSI canceller has a positive effect on the performance of the system in the presence of jamming. In order to evaluate the significance of the DSI canceller in the presence of jamming, the performance of the FM PR is tested across different cancellation parameters. As such, each simulation is passed through the GOCA CFAR filter (described in Table 3.1) using different cancellation parameters and number of successful detections are then counted and compared to the maximum number known to be in the CFAR output under perfect conditions.

Figure 3.8 illustrates the effect of increasing the number of cancellation iterations. The number of cancellations used by the canceller is started at 10, the minimum practical number of iterations required by the CGLS algorithm to allow for convergence to begin. It can be seen from Figure 3.8 that at 10 iterations, even the number of detections when no jamming is present is unsatisfactory.

Increasing the number of iterations to 15 shows improvement in each of the three test metrics, with the greatest improvement seen from the system with no jamming present. Increasing the number of iterations further to 20 resulted in two of the scenarios reaching convergence in terms of maximum detection performance, namely the scenario whereby no jamming is present and where jamming is only present within the reference channel.

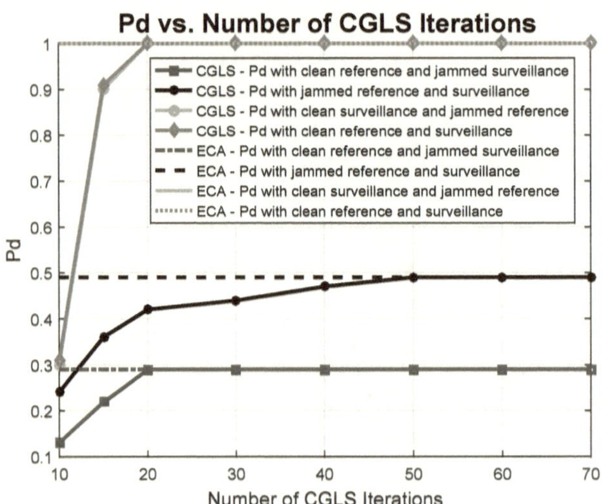

Figure 3.8: Comparing CGLS to ECA cancellation in the presence of jamming. The dashed lines represent the N_d where ECA was used while the solid lines represent N_d when CGLS is applied. The blue lines (squares) represent the scenario where there is a jamming signal present in the surveillance channel but not in the reference channel. The black lines (dots) represent the scenario where there is jamming in both the reference and surveillance channels. The green lines (rings) represent the scenario where jamming is present only in the reference channel but not in the surveillance channel. The red lines (diamonds) represent the scenario where no jamming is present in either channels i.e. the system is clean of any jamming.

Increasing the number of CGLS iterations further to 30 (the number used in all other presented simulations), results in a further increase in the performance of the system with jamming applied to both the reference and surveillance channels, while the other two scenarios remained unchanged in terms of their detection performance. This illustrates very clearly the results shown in equations (3.19) through (3.32). This pattern of detection performance improvement when jamming both the reference and surveillance channels continues until the number of cancellation iterations is increased to 50, at which point convergence is achieved, resulting in all 3 graphs remaining flat regardless of further increases in number of iterations.

A comparison between the two main cancellation techniques is also shown in order to evaluate the performance of each algorithm in the presence of jamming. Figure

3.8 shows the N_d achieved in the presence of jamming when applying ECA or CGLS across various scenarios. It is clear that using ECA results in a significantly more consistent output when compared to using CGLS. This is expected as ECA has a constant output with it being a 'one shot' algorithm where the interference matrix is built and subtracted from the surveillance channel in one go, while CGLS runs for a set number of iterations regardless of whether convergence is achieved or not, refining the cancellation matrix in small increments each iteration [11, 65]. The advantage of ECA is consistency in the output, however the disadvantage is that the execution time and computational load is significantly more than iterative approaches. This makes ECA difficult to implement in real time due to the variability of the FM signal. CGLS however, can be implemented using a smaller fixed matrix size and memory footprint dependent on the number of range and Doppler bins required for cancellation. This leads to a controllable execution time and is therefore suitable for real time implementations.

To elaborate on the effect of the DSI canceller, a derivation of the construction of the ARD maps is shown, beginning with (3.19). The reference signal, $r(t)$, contains the direct path signal as well as some noise, $n_1(t)$. The surveillance signal is therefore made up of some scaled version of the reference signal, $Ar_i(t)$, its own uncorrelated noise, $n_2(t)$, as well as the echo signal, $e(t)$, which is simply a time and Doppler shifted version of the reference signal. The following signal models make the assumption that the jamming signal is at a considerably lower level than the reference signal, as has been the case in the simulations discussed thus far.

$$r(t) = r_i(t) + n_1(t)$$
$$s(t) = Ar_i(t) + e(t) + n_2(t) \tag{3.19}$$

The DSI cancellation algorithm, typically a LMS based algorithm, is used to remove the reference signal found in the surveillance channel, resulting in a new scaling term for the reference signal within the surveillance channel, B, where $B << A$. This results in a new expression for the signal present in the surveillance channel:

$$s_{rm}(t) = Br_i(t) + e(t) + n_2(t) \tag{3.20}$$

where s_{rm} defines the signal in the surveillance channel after DSI cancellation has been applied. Creating the ARD map, χ, is then achieved by correlating the two channels as shown in (3.21).

$$\chi = \int_T r^*(t) s_{rm}(t+\tau) e^{-i2\pi f t} dt \tag{3.21}$$

Introducing the jamming signal results in the jamming signal being imposed on both the reference and surveillance channels, $r_j(t)$ and $s_j(t)$ respectively, provided the receiver channels are co-located as is the case under consideration in (3.22):

$$\begin{aligned} r_j(t) &= r(t) + j_r(t) \\ s_j(t) &= s(t) + j_s(t) \end{aligned} \tag{3.22}$$

where $j_r(t)$ and $j_s(t)$ represent the jamming signal present in the reference and surveillance channels respectively. Ignoring the cancellation step for the time being, the ARD map from the resultant reference and surveillance channels is produced in the same way as before, but now with the added terms.

$$\begin{aligned} \chi_{jr/s}^{\dagger} &= \int_T r_j^*(t) s_j(t+\tau) e^{-i2\pi f t} dt \\ &= \underbrace{\int_T r^*(t) s(t+\tau) e^{-i2\pi f t} dt}_{\chi} + \int_T r^*(t) j_s(t+\tau) e^{-i2\pi f t} dt \\ &\quad + \int_T j_r^*(t) s(t+\tau) e^{-i2\pi f t} dt + \int_T j_r^*(t) j_s(t+\tau) e^{-i2\pi f t} dt \end{aligned} \tag{3.23}$$

The addition of a jamming signal clearly adds additional terms to the ARD map, as seen in the decomposition of $\chi_{jr/s}$ in (3.23). It is shown in (3.22) and (3.23) that the jamming signal appears in both the reference and surveillance channels as an additional interference term. Provided the antennas are co-located, the jamming terms in each channel will be coherent with the only difference being their relative amplitude, i.e., $j_s(t) = C j_r(t)$ where the scaling factor, C, is related to the relative antenna gain in the direction of the jamming signal source.

Applying the same cancellation technique as before, to remove the DSI, provides an interesting result where the jamming signal in the surveillance channel is suppressed slightly by the DSI canceller due to its presence in both channels. The amount of suppression is heavily dependant on the signal level in each channel relative to the reference signal. It is therefore demonstrated that the jamming signal present in the surveillance channel after cancellation is reduced from $j_s(t) \longrightarrow D(\tau, f) j_r(t)$ where $D(\tau, f)$ is a suppression factor representing the cancellation spread in both range and Doppler along with the scaling factor C.

[†]This representation of the ARD map excludes the use of a DSI cancellation scheme.

$$r_j(t) = r(t) + j_r(t)$$
$$s_j(t) = s_{rm}(t) + D(\tau, f)j_r(t) \tag{3.24}$$

Building the ARD map from this gives

$$
\chi_{jr/s} = \underbrace{\int_T r^*(t)s_{rm}(t+\tau)e^{-i2\pi ft}dt}_{\chi}
$$
$$
+ \int_T r^*(t)[D(\tau, f)j_r(t+\tau)]e^{-i2\pi ft}dt
$$
$$
+ \int_T j_r^*(t)s_{rm}(t+\tau)e^{-i2\pi ft}dt \tag{3.25}
$$
$$
+ \int_T j_r^*(t)[D(\tau, f)j_r(t+\tau)]e^{-i2\pi ft}dt
$$

where $\chi_{jr/s}$ represents the output ARD map when jamming is applied to both the reference and surveillance channels. Depending on the amount of suppression achieved through the canceller and therefore the value of $D(\tau, f)$, the amount of jamming signal seen in the output ARD map can vary.

$$
\chi_{j_{r/s}} = \underbrace{\int_T \overbrace{r^*(t)}^{large}\, \overbrace{s_{rm}(t+\tau)}^{small}\, e^{-i2\pi ft}dt}_{\chi}
$$
$$
+ \underbrace{\int_T \overbrace{r^*(t)}^{large}\, \overbrace{[D(\tau, f)j_r(t+\tau)]}^{small}\, e^{-i2\pi ft}dt}_{\chi_{n_1}}
$$
$$
+ \underbrace{\int_T \overbrace{j_r^*(t)}^{small}\, \overbrace{s_{rm}(t+\tau)}^{small}\, e^{-i2\pi ft}dt}_{\chi_{n_2}} \tag{3.26}
$$
$$
+ \underbrace{\int_T \overbrace{j_r^*(t)}^{small}\, \overbrace{[D(\tau, f)j_r(t+\tau)]}^{small}\, e^{-i2\pi ft}dt}_{\chi_{n_3}}
$$

$$\therefore \chi_{j_{r/s}} = \chi + \chi_{n_1} + \chi_{n_2} + \chi_{n_3}$$

For low jamming powers, j_r and s_{rm} are two small uncorrelated signals and the output of their correlation is small compared to that of the other elements, resulting in a very slightly raised noise floor from the χ_{n_2} contribution. This leaves the terms χ_{n_1} and χ_{n_3}. Looking at χ_{n_1}, there are two uncorrelated signals but unlike with χ_{n_2}, where j_r was small, $r(t)$ is large and therefore the output of this correlation can lead to a significant increase in the noise floor of the entire ARD map. χ_{n_3} contains two relatively small signals however, since they are the same, the correlation results in a matched filter response which causes a significant spike in the ARD map around the zero Doppler and zero range bins. The scaling factor $D(\tau, f)$ is important here because it determines the amount of influence each term χ_{n_1} and χ_{n_3} have on the overall ARD map. In the case of χ_{n_3}, almost all of the zero Doppler, zero range interference is removed from the ARD map by the canceller and therefore the term χ_{n_3} will become insignificant, giving us an ARD map consisting of:

$$\therefore \chi_{j_{r/s}} \approx \chi + \chi_{n_1} + \chi_{n_2}$$

The level of interference and increase in noise floor of the ARD map is therefore hugely dependant on χ_{n_1}, which can be reduced by creating a deeper and wider null in the canceller. It is clear that the canceller plays a major role in the performance of the system when both channels are in the presence of jamming. This assumption will only hold true in the case of relatively low jamming power as the other terms in the ARD map will no longer be insignificant.

It must be noted that there are real world limitations to how deep and wide a null can be created with the canceller, namely the ENOB of the ADC in order to capture the full dynamic range of the reference plus echo plus jamming signal, as well as the computational complexity of the canceller for real time output.

Further investigation into the case of jamming only the reference channel or only the surveillance channel is presented. Jamming of only the reference channel results in $j_s(t) = 0$. Applying the DSI cancellation algorithm before creation of the ARD map significantly removes the direct signal, leaving only $s_{rm}(t)$ in the surveillance channel as shown in (3.20). The ARD map can then be constructed as

$$
\begin{aligned}
r_j(t) &= r(t) + j_r(t) \\
s(t) &= s_{rm}(t)
\end{aligned}
\tag{3.27}
$$

$$
\chi_{jr} = \underbrace{\int_T r^*(t) s_{rm}(t+\tau) e^{-i2\pi ft} dt}_{\chi}
\tag{3.28}
$$
$$
+ \underbrace{\int_T j_r^*(t) s_{rm}(t+\tau) e^{-i2\pi ft} dt}_{\chi_{n_2}}
$$

$$\therefore \chi_{jr} = \chi + \chi_{n_2}$$

Because $r(t) >> j_r(t)$ and $j_r(t)$ does not appear in the surveillance channel, the resultant effect is

$$\underbrace{\int_T \overbrace{r^*(t)}^{large}\,\overbrace{s_{rm}(t+\tau)}^{small}\,e^{-i2\pi ft}dt}_{\chi} >> \underbrace{\int_T \overbrace{j_r^*(t)}^{small}\,\overbrace{s_{rm}(t+\tau)}^{small}\,e^{-i2\pi ft}dt}_{\chi_{n_2}} \tag{3.29}$$

As is the case with very low jamming power levels relative to the transmitter power, $\chi_{n_2} \approx 0$ and therefore has very little effect on the overall ARD map output. As the jammer power increases relative to the transmitter power, this assumption no longer holds true. It is clear then that large amounts of jamming power is required to jam through the reference channel alone. This can also be seen by looking at the relative JSR of the reference channel shown in (3.18).

Applying the jamming signal to only the surveillance channel results in a similar ARD calculation but with $j_r(t) = 0$, leading to a vastly different outcome. The DSI cancellation algorithm is applied as usual, leaving $s_{rm}(t) + j_s(t)$ in the surveillance channel. The resultant ARD plot is determined as

$$\begin{aligned}
r(t) &= r(t) \\
s_j(t) &= s_{rm}(t) + j_s(t)
\end{aligned} \tag{3.30}$$

$$\begin{aligned}
\chi_{j_s} &= \underbrace{\int_T r^*(t)s_{rm}(t+\tau)e^{-i2\pi ft}dt}_{\chi} \\
&+ \underbrace{\int_T r^*(t)j_s(t+\tau)e^{-i2\pi ft}dt}_{\chi_{n_1}}
\end{aligned} \tag{3.31}$$

Unlike with jamming only the reference channel, where $j_r(t) << r(t)$, jamming of only the surveillance channel, even with very modest power, leaves $j_s(t) \approx s_{rm}(t)$. Correlating these two signals with the reference signal results in

$$\begin{aligned}
\chi_{j_s} &= \underbrace{\int_T \overbrace{r^*(t)}^{large}\,\overbrace{s_{rm}(t+\tau)}^{small}\,e^{-i2\pi ft}dt}_{\chi} \\
&+ \underbrace{\int_T \overbrace{r^*(t)}^{large}\,\overbrace{j_s(t+\tau)}^{small}\,e^{-i2\pi ft}dt}_{\chi_{n_4}}
\end{aligned} \tag{3.32}$$

$$\therefore \chi_{j_s} = \chi + \chi_{n_4}$$

Comparing all three scenarios for the ARD map creation, it is clear that with the addition of DSI cancellation, an FM PR will perform better if the jamming signal is present in both the reference and surveillance channels. The best performance in the presence of jamming is achieved when the surveillance channel is clear of the jamming signal i.e., the jamming signal is applied to the reference channel only. This is generally only a consideration with spatially separated reference and surveillance channels.

3.7 Chapter Summary

This chapter has presented an overview of the FM signal structure along with a typical FM PR processing chain. Simulations have been performed to determine the effect of a DSI canceller on the performance of an FM PR in the presence of jamming. This was achieved by applying a noise jamming signal to each of the radar channels, first investigating jamming only the reference channel while maintaining a clean surveillance channel and then vice versa, effectively simulating a system utilising a separated reference and surveillance channel. Jamming was then applied to both the reference and surveillance channels as would be the case of a system with co-located reference and surveillance channels.

The results demonstrated in Figure 3.8 indicate that an FM PR is more vulnerable to jamming when using separated reference and surveillance channels. Jamming only the reference channel results in the least system performance degradation while jamming only the surveillance channel results in the most performance degradation, more so than when both channels are jammed ($N_d = 29\%$ vs $N_d = 44\%$). It was shown that this result is due to the ability of the DSI canceller to partially suppress the jamming signal when it appears in both reference and surveillance channels.

Chapter 4

FM Jammer Waveform Design

The results shown in Section 3.6 illustrate the importance of the DSI cancellation stage of the FM radio based PR processing chain, not only in normal operation but also in the presence of noise jamming. The effectiveness of noise jamming against an FM radio based PR when the location of the receiver is known has also been demonstrated.

This chapter expands on the concepts demonstrated through noise jamming and investigates the effect of using more advanced, tone based and Wide Band FM (WBFM) modulated waveforms broadcast on the same FM frequency channel as the reference signal in order to determine the optimal jammer waveform. The effects of message signal bandwidth on the jammer performance has also been investigated by varying the modulation index, β, of the jamming signal.

The same simulation procedure that was followed in Section 3.6 was performed again, this time to determine the effects of different jamming waveforms on the representative FM PR. Table 3.1 once again summarises the simulation parameters used with each simulation. As demonstrated previously, a clean simulation was performed each time without jamming present to provide a benchmark for system performance.

The waveforms investigated include:

- Broadband AWGN jamming

- Single tone jamming on carrier and pilot tones

- Broadband AWGN jamming modulated using WBFM modulation ($\beta = 5$)

- High bandwidth message signal modulated with WBFM modulation ($\beta = 4$)

- Medium bandwidth message signal modulated using WBFM modulation ($\beta = 2$)

- Low bandwidth message signal modulated using WBFM modulation ($\beta = 0.25$)

One of the more difficult to quantify aspects of FM PR is the specific system performance for a given target with a given set of parameters. This difficulty is due to the reference FM waveform itself in that the signal structure and bandwidth fluctuates based on the on-air content over the CPI. This uncontrollable trait of the FM waveform results in largely non-deterministic system performance from CPI to CPI. It is therefore difficult to quantify jamming performance in a generic way. To circumvent this problem, the effectiveness of the attack is assessed on the level of the CFAR output across multiple simulations utilising different transmit waveforms for statistical diversity.

To build up a basic statistical profile of each jamming waveform, each scenario was simulated multiple times, each time incrementing the JSR_E, starting at 30 dB and going as high as 70 dB. The JSR_E represents the ratio of the jammer power to the target echo power at the input terminals of the receiver antenna as described in Section 3.3. The number of detections at the output of the CFAR was then compared to the expected number from a clean simulation where no jamming was present. Due to the unpredictable nature of an FM waveform, this process was repeated 4 times, each time using a different transmit waveform to allow for signal variations and outlying anomalies. This process allows for proper quantification of the effects of each jammer waveform. The simulation parameters used are summarised in Table 3.1.

4.1 Broadband AWGN Jamming

As demonstrated in Section 3.6, AWGN jamming has the effect of increasing system noise such that the overall noise floor at the output ARD map was increased, potentially masking the target. Using the CFAR parameters described in Table 3.1, figure 4.1 summarises the simulation output for each jamming scenario with the y-axis representing the number of detections normalised to a clean simulation with no jamming and the x-axis representing the applied JSR.

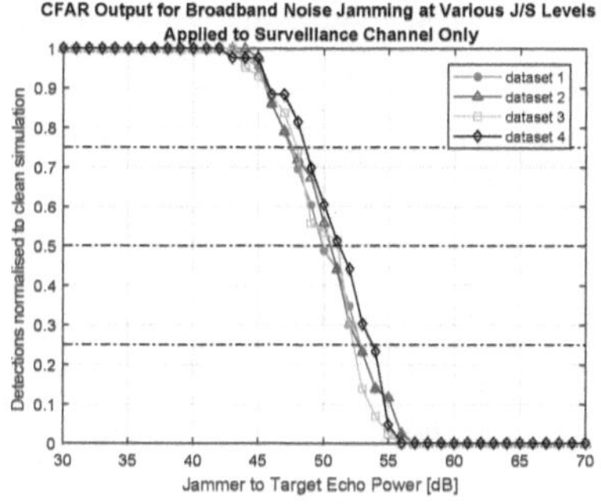

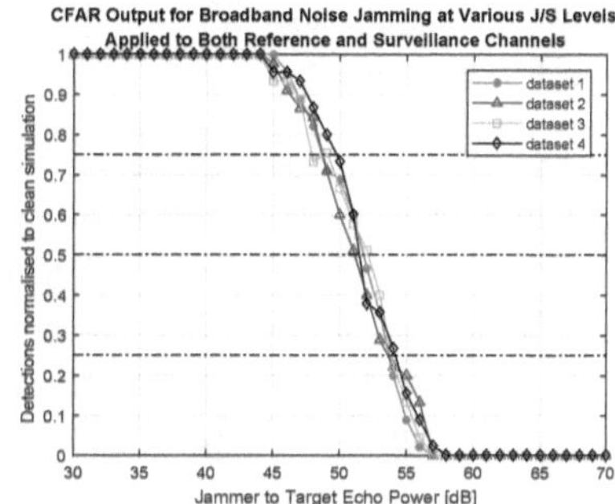

(a) Jamming applied to surveillance channel only

(b) Jamming applied to Reference and Surveillance channels

Figure 4.1: System performance with AWGN as an attack signal.

Comparing Figures 4.1a and 4.1b it is clear that noise jamming has a relatively consistent effect on the performance of the system, regardless of the on-air content transmitted. Averaging the outputs across each JSR_E value produces the results demonstrated in Figure 4.2.

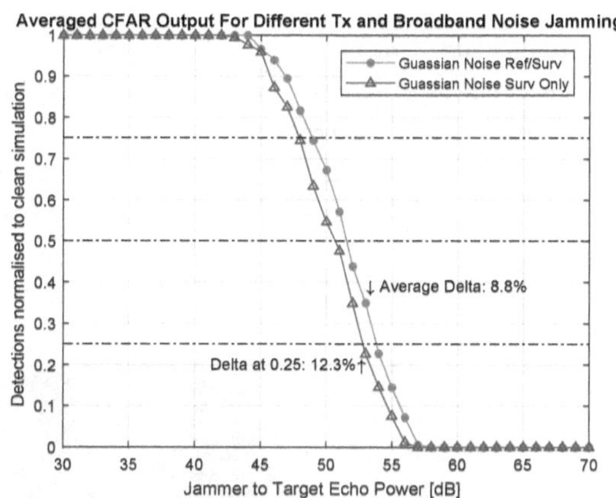

Figure 4.2: Performance comparison between systems with broadband noise jamming applied. (Red) Represents when jamming was applied to both the reference and surveillance channels. (Blue) Represents when jamming was applied to only the surveillance channel.

The red dots represent the PR performance when the jamming signal appears in both the reference and surveillance channels while the blue triangles represent the PR performance when the jamming signal appears only in the surveillance channel. As shown in Chapter 3.6, there is an improvement in the PR performance for the same JSR_E

when the jamming signal appears in both the reference and surveillance channels due to the DSI canceller.

The average performance difference was assessed by comparing the number of normalised detections of each graph for each JSR_E, excluding the end points where the differential was less than 1%. As demonstrated in Figure 4.2, the average difference was an 8.8% improvement in the number of positive detections when 30 cancellation iterations were used. The ability to differentiate between the target and false detections within the CFAR output diminishes drastically below the 25% detection point.

Comparing each scenario at the 25% detection point, the number of detections increases by 12.3% from 20% to 32.3% when the jamming signal appears in both channels. The JSR_E required to reduce the number of detections below the 25% point for each scenario was 52 and 53 dB respectively. This indicates a 1-2 dB increase in required JSR_E or, as previously adumbrated, an average improvement of 8.8% in overall number of detections as a result of the canceller.

4.2 Single Tone Jamming

The following simulation results demonstrate the effect on the system when jamming was applied using a single tone imposed onto the carrier frequency. Figure 4.3a demonstrates the output of the single tone jamming simulation where peaks can be seen to appear across the entire range profile at various Doppler frequencies where the target is shown in the red oval. Unlike when Gaussian noise jamming was used where the overall noise floor was raised, jamming with a single tone on the carrier creates artefacts in the output ARD map that were easy to discern as intentional jamming.

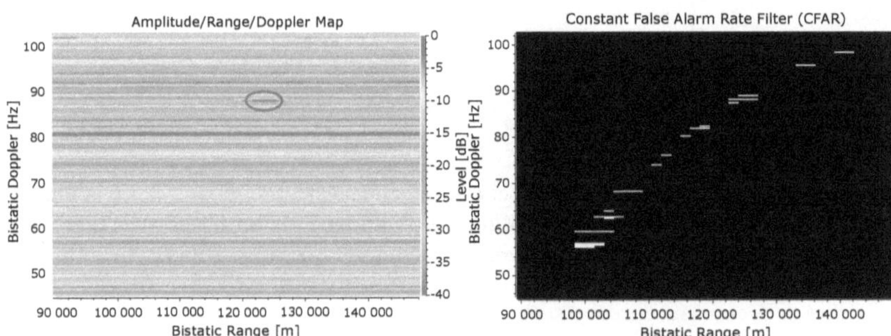

(a) ARD plot where target is shown in the red oval

(b) Combined CFAR output, applied in the range dimension

Figure 4.3: System performance for simulated tone jamming on carrier.

Filtering the output ARD maps through the CFAR filter (applied in the Doppler dimension as described in Table 3.1), results in zero target detections along the simulated flight track. Applying the CFAR filter in the range dimension rather than the Doppler dimension results in partial detections ($N_d = 35\%$) along the flight path as demonstrated in Figure 4.3b. The disadvantage of using a CFAR filter operating in the range dimension for an FM PR is its sensitivity to bandwidth fluctuations.

The CFAR filter was designed to provide optimum detection performance based on the expected background noise which, in the case of an FM radio waveform, was assumed to be Gaussian. The high sidelobe peaks throughout the range profile results in the background noise profile assumption to be incorrect, therefore significantly reducing detection performance. To counter this, the CFAR filter was applied in the range dimension instead, however, this comes with its own drawbacks. It is therefore suggested that when interference of this nature is detected, a dynamically selected CFAR dimension can be used. This way, under normal operation, a CFAR applied in the Doppler dimension can be used to maintain its robustness and a range based CFAR filter can be used when jamming or unintentional interference is detected. The range based CFAR detector, under these conditions, detects only the target with zero false detections. This could assist the tracking filter in achieving and maintaining a track since there were no false targets.

As with the broadband AWGN, Figures 4.4a and 4.4b summarise the performance of separated reference and surveillance and co-located reference and surveillance channels in the presence of single tone jamming over various $\mathrm{JSR_E}$ when the CFAR filter was applied in the Doppler dimension.

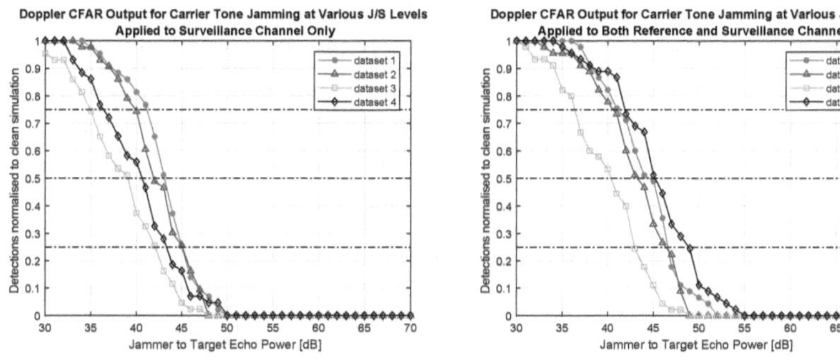

(a) Jamming applied to surveillance channel only

(b) Jamming applied to Reference and Surveillance channels

Figure 4.4: System performance with tone jamming on carrier (Doppler CFAR).

As the bandwidth of the reference signal decreases, the carrier tone and pilot tone levels increase as illustrated in Section 3.1. This leads to an increase in jamming effectiveness due to the increased sidelobe peaks as illustrated in Figures 4.4a and 4.4b where the lowest bandwidth reference signal, dataset 3 (green squares), demonstrates the largest deterioration due to jamming. Averaging the outputs across each JSR_E produces the results demonstrated in Figure 4.5.

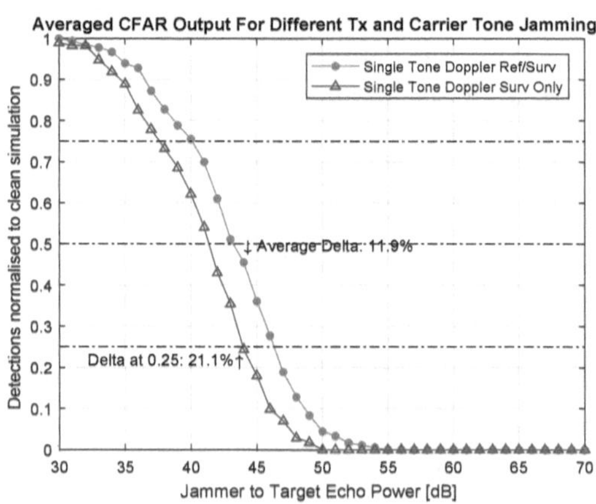

Figure 4.5: Performance comparison between systems where carrier tone jamming has been applied with a Doppler based CFAR detector. (Red) Represents when jamming was applied to both the reference and surveillance channels. (Blue) Represents when jamming was applied to only the surveillance channel.

The average performance difference between the two system set-ups was 11.9% while the performance difference at the 25% mark was 21.1%. The JSR_E required to reduce the number of detections to below the 25% point for each scenario was 44 and 47 dB, meaning that the JSR_E needs to be 2-3 dB higher for the co-located system to achieve the same degradation in performance as the separated system as a result of the DSI canceller when applying the CFAR in the Doppler dimension.

The results of applying the CFAR filter in the range dimension are illustrated in Figures 4.6a and 4.6b. While the performance in the presence of tone jamming was improved, the unpredictable nature of range based CFAR applied to FM PR due to the bandwidth and subsequent range resolution fluctuations was clear.

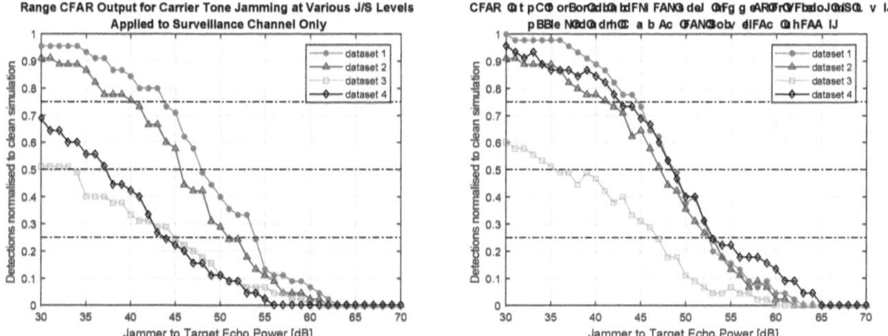

(a) Jamming applied to surveillance channel only   (b) Jamming applied to Reference and Surveillance channels

Figure 4.6: System performance with tone jamming on carrier (Range CFAR).

Averaging the outputs of Figure 4.6 across each JSR_E produces the results demonstrated in Figure 4.7.

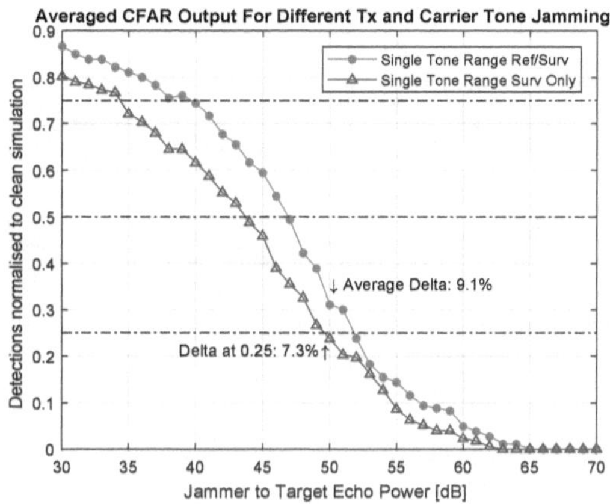

Figure 4.7: Performance comparison between systems where carrier tone jamming has been applied with a range based CFAR detector. (Red) Represents when jamming was applied to both the reference and surveillance channels. (Blue) Represents when jamming was applied to only the surveillance channel.

From Figure 4.7, the average difference between the two systems was 9.1% while the performance difference at the 25% point was 7.3%. The JSR_E required to reduce the number of detections below the 25% point for each of the scenarios was 50 and 52 dB respectively, indicating a 1-2 dB increase in required JSR_E in order to achieve the same performance degradation when jamming both channels due to the DSI canceller.

Comparing the Doppler CFAR output to range CFAR output demonstrates that the range CFAR was more consistent at higher levels of JSR_E, while the Doppler CFAR outperforms the range CFAR when low power or no tone jamming was present. The downside to implementing a range CFAR for FM PR is, however, also clear from Figure 4.6 where a low bandwidth reference signal, (dataset 3 - green squares), results in poor range resolution. This causes the CFAR filter to break down with only 62% of the total detections being made when no jamming was applied.

4.3 Jamming using WBFM

Additive white Gaussian noise has been shown to have the effect of raising the noise floor while tones can be used to produce ridges throughout the range dimension of the ARD map. To achieve any significant correlation between reference and jamming waveforms, the deterministic components of the FM waveform need to be exploited. As discussed in Section 3.1, there are four deterministic components with respect to a commercial FM waveform: the carrier tone, the stereo tone, the left-right tone and the RDS data stream.

To determine the performance of the system when the deterministic components of the transmitted FM waveform are attacked, the following WBFM jamming waveforms were used:

- AWGN modulated as a WBFM waveform ($\beta = 5$)

- High bandwidth message signal modulated as a WBFM waveform ($\beta = 4$)

- Medium bandwidth message signal modulated as a WBFM waveform ($\beta = 2$)

- Low bandwidth message signal modulated as a WBFM waveform ($\beta = 0.25$)

4.3.1 Noise Modulated FM Jammer ($\beta = 5$)

To create the noise modulated FM waveform, WBFM modulation was applied to a baseband Gaussian white noise message signal, such as the one used in the Gaussian white noise simulations, giving the jammer waveform a constant β value. The output of the WBFM modulation was then an FM signal that when demodulated, was Gaussian white noise.

As with unmodulated broadband Gaussian white noise, the WBFM modulated noise causes increasing performance degradation with increasing $\mathrm{JSR_E}$, however, at significantly lower $\mathrm{JSR_E}$ levels as demonstrated in Figures 4.8a and 4.8b.

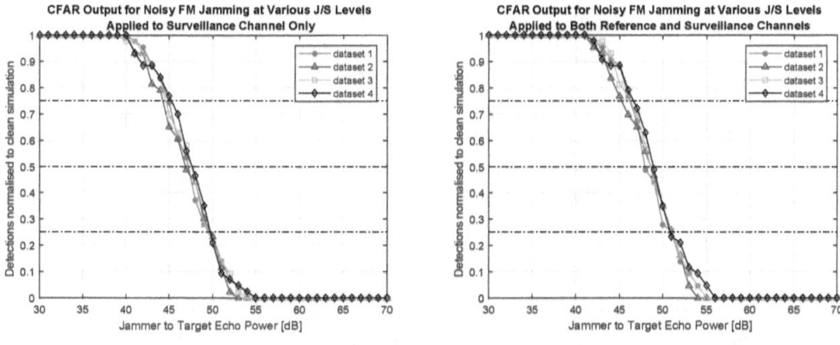

(a) Jamming applied to surveillance channel only

(b) Jamming applied to Reference and Surveillance channels

Figure 4.8: System performance with high bandwidth FM jamming.

Averaging the outputs across each $\mathrm{JSR_E}$ in Figures 4.8a and 4.8b produces the results demonstrated in Figure 4.9.

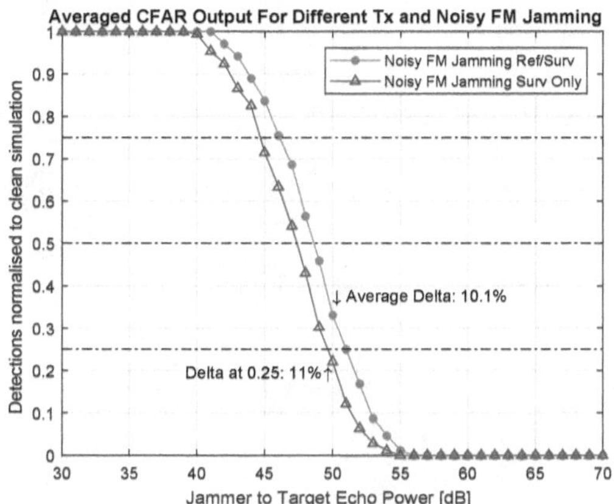

Figure 4.9: Averaged CFAR performance comparison between systems where noise modulated FM jamming has been applied. (Red) Represents when jamming was applied to both the reference and surveillance channels. (Blue) Represents when jamming was applied to only the surveillance channel.

Once again it was clear that, due to the DSI canceller when the jamming signal appears in both the reference and surveillance channels, the average performance improvement was 10.1% compared to when the jamming signal only appears in the surveillance channel. A 3 dB reduction in required JSR_E was observed when WBFM modulation was applied to the noise signal as seen when comparing the system performance using modulated and unmodulated noise in Figures 4.9 and 4.2 respectively.

4.3.2   High Bandwidth FM Jammer ($\beta = 4$)

Increased tone correlation, as demonstrated with tone jamming, can be achieved by increasing the tone levels in the jamming signal. Increasing the tone levels was achieved by decreasing the bandwidth of the message signal at the input. To illustrate the effect of increasing the tone levels, simulations were run where the bandwidth of the jammer message signal, $m(t)$, was varied. The results of a high bandwidth message signal ($\beta = 4$).

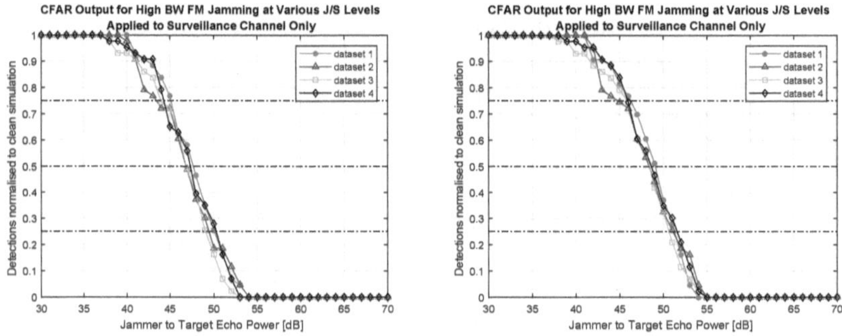

(a) Jamming applied to surveillance channel only

(b) Jamming applied to Reference and Surveillance channels

Figure 4.10: System performance with high bandwidth FM jamming.

Averaging the outputs across each JSR_E in Figures 4.10a and 4.10b produces the results demonstrated in Figure 4.11.

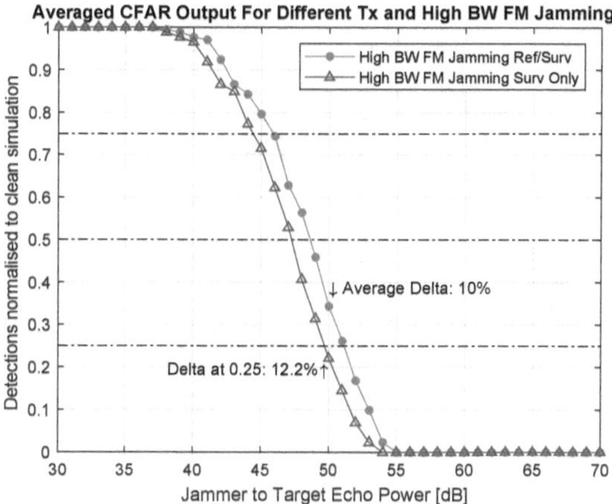

Figure 4.11: Averaged CFAR performance comparison between systems where high bandwidth music FM jamming has been applied. (Red) Represents when jamming was applied to both the reference and surveillance channels. (Blue) Represents when jamming was applied to only the surveillance channel.

The WBFM jamming signals with similar bandwidths produce similar jamming performance, as illustrated in Figures 4.9 and 4.11 where both jamming signals have similar β values. The JSR_E required to reduce the relative PR system performance by 75% was 50 and 51 dB when applied to only the surveillance and then the reference and surveillance respectively, for both the noise modulated FM and high bandwidth FM jamming.

High bandwidth FM jamming performs marginally better at lower JSR_E, and was almost identical to noise modulated FM jamming as the JSR_E increases. This implies that the slightly lower β value found in the high bandwidth FM jammer waveform results in slightly higher pilot levels and therefore marginally increased integration gain. This indicates that the content itself was insignificant to the performance of the jammer, and rather the β value of the resultant message signal, $m(t)$, that has the biggest influence on the jamming performance of a WBFM jamming signal.

4.3.3 Medium Bandwidth FM Jammer ($\beta = 2$)

Reducing the β value of the modulated FM jamming signal from $\beta = 4$ to $\beta = 2$ results in an increase in the carrier tone levels. To demonstrate the effect of lowering β, a WBFM was applied to a medium bandwidth message signal ($\beta = 2$) as demonstrated in Figure 4.12 and used as the jamming signal.

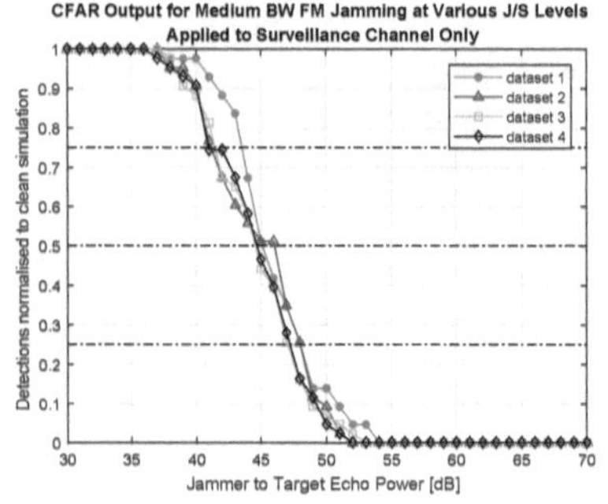
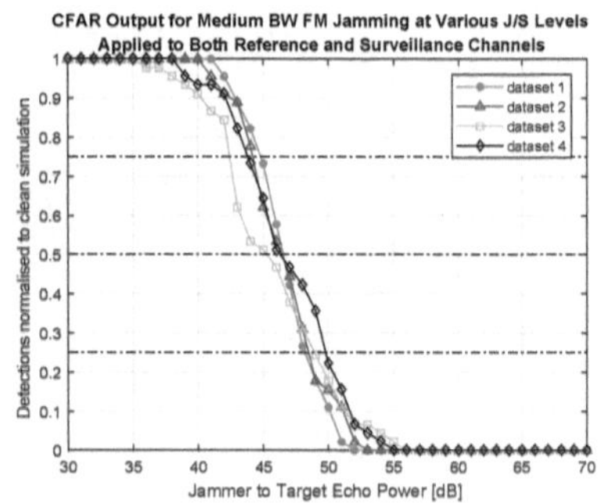

(a) Jamming applied to surveillance channel only

(b) Jamming applied to Reference and
Surveillance channels

Figure 4.12: System performance with medium bandwidth FM jamming.

Averaging the outputs across each JSR_E produces the results demonstrated in Figure
4.13.

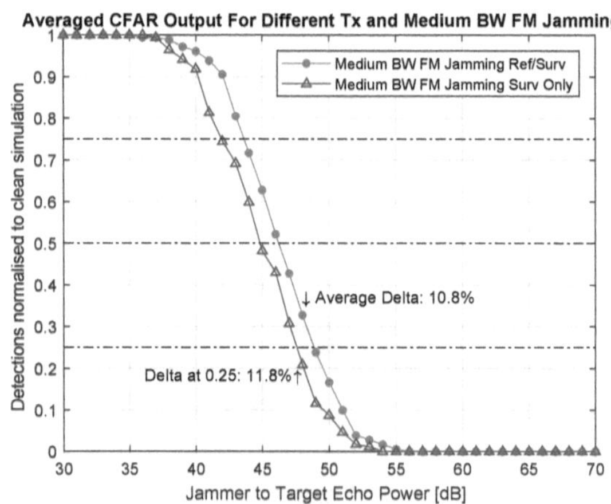

Figure 4.13: Performance comparison between systems with medium bandwidth music FM jamming being applied. (Red) Represents when jamming was applied to both
the reference and surveillance channels. (Blue) Represents when jamming was applied to only the surveillance channel.

The results demonstrated in Figure 4.13 illustrate the effect of using a jamming signal
with a lower β value, i.e. with medium bandwidth. Even at low power levels, the ability
of the system to detect a target was greatly diminished. The average performance
difference between the two system set-ups was 10.8%. The JSR_E required to reduce

the number of detections to below the 25% point for each scenario was 48 and 49 dB, meaning that the JSR_E needs to be 1-2 dB higher for the co-located system to achieve the same degradation in performance as the separated system.

4.3.4   Low Bandwidth FM Jammer ($\beta = 0.25$)

Decreasing the bandwidth further to $\beta = 0.25$, results in a significant increase in the tone levels within the jamming waveform. Figure 4.14 illustrates the output of simulations using a WBFM jamming waveform with a $\beta = 0.25$.

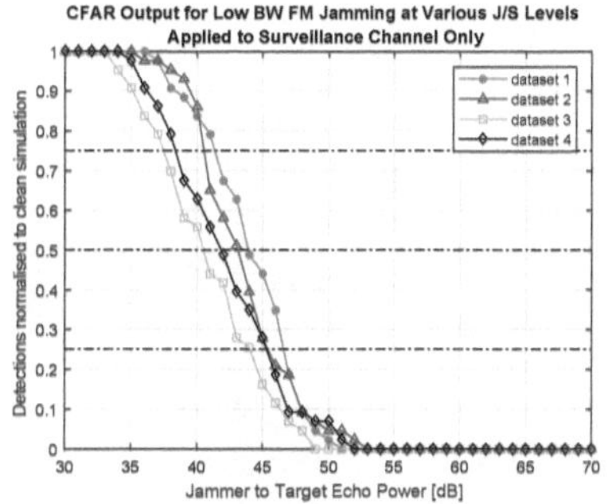
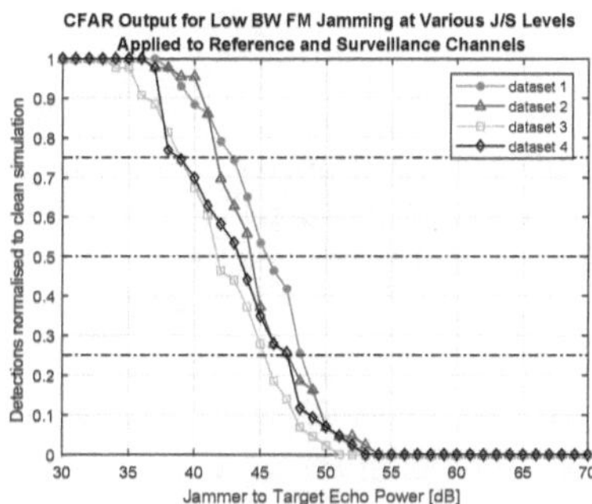

(a) Jamming applied to surveillance channel only

(b) Jamming applied to Reference and Surveillance channels

Figure 4.14: System performance with low bandwidth FM jamming

Averaging the outputs across each JSR_E produces the results demonstrated in Figure 4.15.

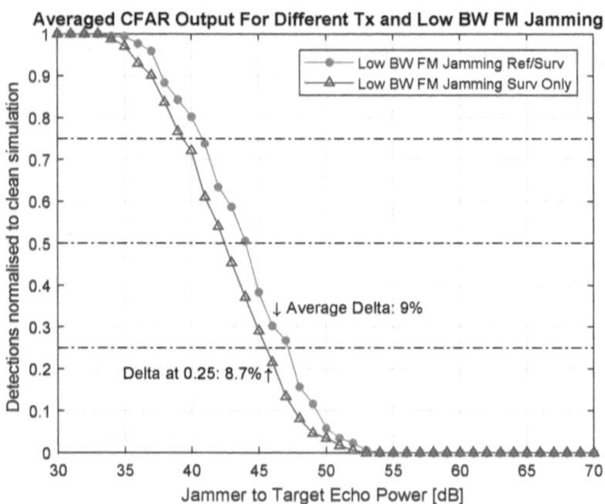

Figure 4.15: Performance comparison between systems with low bandwidth music FM jamming being applied. (Red) Represents when jamming was applied to both the reference and surveillance channels. (Blue) Represents when jamming was applied to only the surveillance channel.

As with all the simulations, applying the jamming waveform in only the surveillance channel results in the greatest reduction in PR performance. The average difference in performance between the two system set-ups was 9%. Comparing the results from Figure 4.15 to the results demonstrated in Figures 4.9, 4.11 and 4.13, it is clear that when using a WBFM jamming waveform, the relative system performance decreases with decreasing β.

4.4 Discussion of Results

To evaluate whether the FM PR was jammed or rendered ineffective, a base level of acceptable performance needs to be defined. For the purpose of this study, a system was defined as effectively jammed when its detection performance was below 25%. To compare the performance of each jamming waveform against each other, the results of each simulation were plotted on a single set of axis as demonstrated in Figures 4.16 and 4.17 where jamming applied to the surveillance channel only and then to both the reference and surveillance channels respectively.

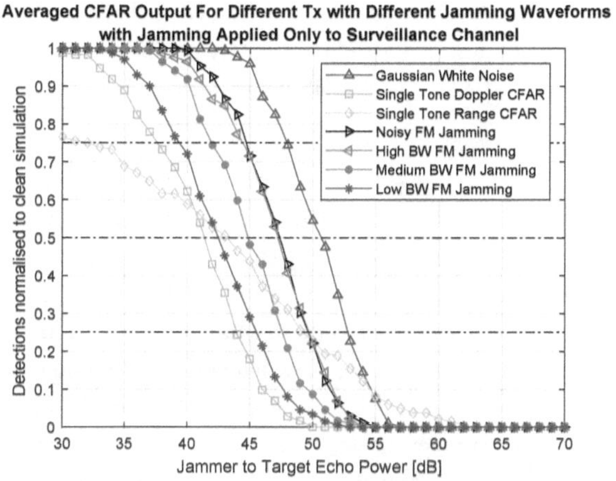

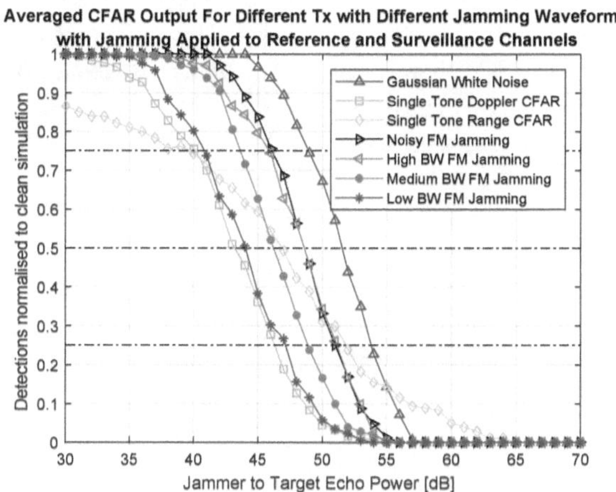

Figure 4.16: Combined averaged plots for each jamming waveform across various JSR_E when applied in only the surveillance channel.

Figure 4.17: Combined averaged plots for each jamming waveform across various JSR_E when applied to both the reference and surveillance channels.

Analysing the results from Figures 4.16 and 4.17 illustrates that the most effective form of jamming an FM PR was when the deterministic components of the FM waveform were attacked. Noise jamming has a consistent and predictable effect on the system since the receiver becomes more and more desensitised as the JSR_E increases. The same performance degradation curve was seen when using high bandwidth WBFM jamming, however, the curve itself was shifted to the left indicating that lower jamming power

66

was required to achieve the same level of effectiveness e.g., 56 dB JSR_E was required to reduce the number of detections to below 25% when using noise jamming but only 51 dB JSR_E was required when using high bandwidth WBFM jamming.

This result demonstrates that the message signal content itself, $m(t)$, was insignificant to the performance of the jammer, and rather it was the β value of the modulated signal that plays the dominant role. This was due to the increase in carrier tone levels as the bandwidth, and subsequently β decreases while the average power of the transmitted spectrum remains constant as defined by (3.10) in Section 3.1. The closer the waveform tends towards absolute tone jamming, the more the sidelobe peaks become apparent in the output ARD maps as demonstrated in Figures 4.18(a-e). While the reference message signal bandwidth can not be controlled by a jammer operator, a combination of both low jammer message signal bandwidth and low reference message signal bandwidth and high bandwidth fluctuations across a single CPI leads to to improved jamming effectiveness.

Evaluating the background noise in the output ARD for each jamming waveform in Figures 4.18(b-f) compared to when no jamming is applied in Figure 4.18(a), illustrates that as β decreases, the background noise of the ARDs becomes less Gaussian. The SNR of the target in each plot was significantly impacted, going from 42.2 dB when no jamming was present to 13.2 dB with $\beta = 5$, 11.7 dB with $\beta = 4$, 12.9 dB with $\beta = 2$ and 11.3 dB with $\beta = 0.25$ (when $JSR_E = 50$ dB).

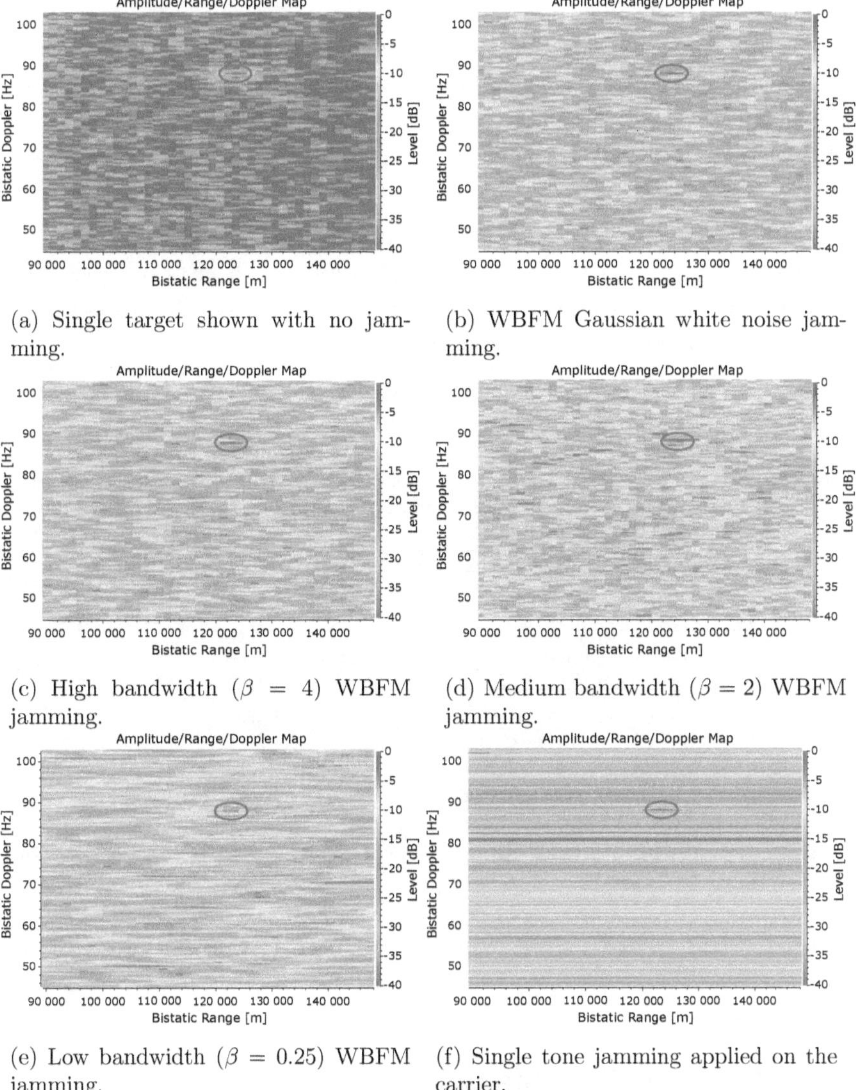

(a) Single target shown with no jamming.

(b) WBFM Gaussian white noise jamming.

(c) High bandwidth ($\beta = 4$) WBFM jamming.

(d) Medium bandwidth ($\beta = 2$) WBFM jamming.

(e) Low bandwidth ($\beta = 0.25$) WBFM jamming.

(f) Single tone jamming applied on the carrier.

Figure 4.18: ARD output for each jamming waveform for simulations where $\mathrm{JSR_E} = 50$ dB, compared to the clean simulation, with target highlighted in red oval.

The effect of carrier tone jamming on the CFAR output is severe. Carrier tone jamming causes large sidelobe peaks to appear throughout the Doppler profile of the ARD plot. This in turn, causes a target that appears near these peaks to be masked, leading to a dramatic reduction in CFAR detection performance. The reduction in CFAR performance was due to the CFAR filter operating on the assumption that the ARD background noise was Gaussian due to the Gaussian-like properties of the FM waveform. As a result, when carrier tone jamming was applied, the high peaks and low troughs cause the background reference cells to behave in a non-Gaussian fashion, severely degrading the detection performance of the filter.

Table 4.1 summarises the results of using different jamming waveforms where the jamming was applied to both the reference and surveillance channels and when it was applied to only the surveillance channel. The relative performance differences between the two scenarios were also indicated in the table.

Table 4.1: Summary of the required $\mathrm{JSR_E}$ for a desired reduction in system performance* given a specific jamming waveform.

Number of CGLS iterations	Jammer Waveform	$\mathrm{J/S_E}$ required to reduce detection performance by X% referenced to the clean simulation					
Iterations	**Gaussian White Noise**	25%	50%	75%	100%	**Delta at 25%**	**Ave Delta**
30	Surveillance Only	48 dB	51 dB	53 dB	57 dB	12.3%	8.8%
30	Reference and Surveillance	49 dB	52 dB	54 dB	58 dB		
Iterations	**Noise Modulated FM ($\beta = 5$)**	25%	50%	75%	100%	**Delta at 25%**	**Ave Delta**
30	Surveillance Only	45 dB	48 dB	50 dB	55 dB	11%	10.1%
30	Reference and Surveillance	46 dB	49 dB	51 dB	56 dB		
Iterations	**High BW FM ($\beta = 4$)**	25%	50%	75%	100%	**Delta at 25%**	**Ave Delta**
30	Surveillance Only	45 dB	48 dB	50 dB	54 dB	12.2%	10%
30	Reference and Surveillance	46 dB	49 dB	51 dB	55 dB		
Iterations	**Medium BW FM ($\beta = 2$)**	25%	50%	75%	100%	**Delta at 25%**	**Ave Delta**
30	Surveillance Only	42 dB	45 dB	48 dB	54 dB	11.8%	10.8%
30	Reference and Surveillance	44 dB	46 dB	49 dB	55 dB		
Iterations	**Low BW FM ($\beta = 0.25$)**	25%	50%	75%	100%	**Delta at 25%**	**Ave Delta**
30	Surveillance Only	40 dB	43 dB	46 dB	53 dB	8.7%	9%
30	Reference and Surveillance	41 dB	44 dB	48 dB	54 dB		
Iterations	**Carrier Tone (Doppler CFAR)**	25%	50%	75%	100%	**Delta at 25%**	**Ave Delta**
30	Surveillance Only	38 dB	42 dB	44 dB	50 dB	21.1%	11.9%
30	Reference and Surveillance	40 dB	44 dB	47 dB	55 dB		
Iterations	**Carrier Tone (Range CFAR)**	25%	50%	75%	100%	**Delta at 25%**	**Ave Delta**
30	Surveillance Only	37 dB	45 dB	50 dB	63 dB	5.8%	6.3%
30	Reference and Surveillance	40 dB	47 dB	52 dB	65 dB		

* The system performance indicated here is representative of the system as described in Table 3.1 where jamming was applied to both the reference and surveillance channels and 30 iterations of CGLS DSI cancellation was used.

4.5   Chapter Summary

The effects of various jamming waveforms applied to FM PR have been investigated and the results have been summarised in Table 4.1. It has been shown that the most effective jamming waveform was a single tone transmitted on the same centre frequency as the reference signal being used by the FM PR. It has also been demonstrated that when the jamming signal appears in both the reference and surveillance channels, the DSI canceller can improve the performance of an FM PR by approximately 10%

on average when compared to a system where the jamming signal appears in only the surveillance channel. The importance of knowledge of the receiver location is demonstrated in Appendix A where the required jamming power for different scenarios is demonstrated.

Since FM waveforms have a constant average power, approximated by (3.10), by decreasing the number of sidebands required to accurately represent the message signal, (decreasing β), the carrier tone levels increase as demonstrated in Figure 3.1. As the jamming signal becomes more tone-like, severe artefacts start to appear across the range profile in the ARD map output. Applying a CFAR filter to the ARD map in the Doppler domain results in complete masking of the target as the peaks lead to a change in the background noise model of the system which causes the CFAR filter assumptions to break down, resulting in incorrect threshold levels, further degrading system performance. Applying the CFAR filter in the range dimension improves the performance slightly from 0% detections to 35% detections relative to the case when no jamming is applied. It is clear from these results that an appropriate ES receiver is required in order to detect the presence of jamming and for the PR to put measures in place to try counter any potential jamming. One possible counter measure would be to dynamically switch to applying the CFAR filter in the range dimension when single tone jamming is detected. Another could be to dynamically increase the depth of the cancellation null to attempt to further suppress the jamming signal. Monitoring the channel impulse response (zero-Doppler line) in an attempt to locate additional peaks caused by the jammer could be implemented. This could possibly also assist in detecting the presence of noise-jamming.

For more advanced receiver systems, tone blanking or suppression could potentially be used as an effective counter to tone jamming on any of the pilot tones. This would be accomplished by removing the pilot tones from the surveillance channel before processing. This would not, however, counter tones that were broadcast at arbitrary frequencies within the FM channel spectrum or pilots whose spectral content is rapidly fluctuating as is often the case. For this, a new CFAR filter will be required that takes into account the change in the statistical profile of the background noise as a result of the jamming.

Chapter 5

Measured FM Passive Radar Results

To verify the results of the simulations in Chapter 4, measurements were made using a real PR at UCT. The system used for these measurements was the ComRAD3 PR radar system by Peralex [113]. The ComRAD3 system is a 3 channel direct conversion wideband receiver that covers the entire FM band from 88 MHz to 108 MHz.

The measured results illustrate the performance of the two most contrasting simulated jamming waveforms, wide-band noise jamming and a single tone on the carrier. Due to the uncontrollable nature of the FM transmit waveform, demonstrating the two edge cases provides the most repeatable and deterministic results to compare with the simulations shown in Chapter 4.

5.1 System set-up

The system specifications used for the field measurements were identical to those used in the simulations with the exception of the basic system geometry, centre frequency, cancellation bin size, clutter model, target size and Swerling model. The complete system parameters for the measurements can be found in Table 5.2 while the parameters that differ from the ones used in the simulations are outlined in Table 5.1.

Table 5.1: Differences between FERS simulated system parameters and the measured system parameters.

Parameters	Simulation	Measurement
Carrier Frequency	89 MHz	91.1 MHz
Cancellation Bins	5 range, 5 Doppler	120 range, 5 Doppler
Clutter	None	Environmental
Target	23 dBsm, Swerling 0	Boeing 737-800

It must be noted that the difference between the simulated and measured cancellation bins (5 vs. 120) is due to the lack of simulated clutter. This means that only the direct signal i.e. zero range and Doppler bins, were required to be cancelled and increasing the number of range bins past the first few achieve little more than increase the processing overhead. Figure 5.2 shows the antenna set-up used for the field trials. It can be seen that there were a total of 3 antennas, two surveillance antennas situated below a reference antenna. The reference and surveillance antennas were perpendicular to each other with the reference antenna pointing towards the Piketberg transmitter [149] and the surveillance antennas pointed towards the Cape Town International airport with a slight upwards tilt. The complete system geometry is demonstrated in Figure 5.1 where the ADS-B tracks of the three detected targets can be seen. This is the same system set-up that was used in [94].

The transmitter to receiver baseline distance was 118.5 km between the UCT receiver site and the Piketberg transmit site. The Piketberg transmitter, which transmits FM radio using a vertically polarised dipole antenna with channel center frequencies of 88.0 MHz, 91.1 MHz, 94.3 MHz, 97.6 MHz, 101.1 MHz, 104.7 MHz and 107.6 MHz, each with an ERP of 10 kW. Further details regarding the transmitter can be found in [149]. It was observed that 91.1 MHz exhibited higher average bandwidth and better channel separation than the other channels and was therefore chosen as the channel of choice for the field trials.

The power difference between the reference and surveillance channels ranged between 7 dB and 15 dB with the average difference typically above 10 dB. This separation was achieved by stacking the reference and surveillance antennas vertically and per- pendicularly to each other as demonstrated in Figure 5.2.

While the ComRAD3 system is capable of digitising 3 channels coherently (two surveil- lance and one reference, as evident by the 3 antennas), only a single reference and single surveillance antenna were used for these trials. Channel 1 was used as the reference channel and channel 3 as the surveillance channels. For these field trials, channels 1 and 3 were used as they offered the greatest channel separation. Channel 2 was not used.

Table 5.2: System parameters used for measurement campaign.

Transmitter (Tx)	
Antenna Beam Pattern	Isotropic
Antenna Gain	2.15 dBi
Antenna Altitude	850 m
Carrier Frequency	91.1 MHz
ERP	10 kW
Waveform	Commercial FM radio
Reference and Surveillance Receivers (Rx)	
Antenna Beam Pattern	Sinc
Antenna Gain	7.2 dBi
Antenna Altitude	140 m
LO Error	50 ppb (std. dev. of 0.01 Hz @ 204.8 kSps)
Noise Figure	4 dB
Digitisation	204.8 kSps complex, 16 bit quantisation
Tx to Rx Baseline	118 500 m
Target	
Initial Altitude	NA
Final Altitude	NA
Velocity	100 - 300 m/s
RCS @ 91.1 MHz	Unknown
Jammer	
Antenna Beam Pattern	isotropic
Antenna Gain	2.15 dBi
Transmit Power	-35 dBm before antenna gain
Carrier Frequency	91.1 MHz
Waveform	AWGN, Sine wave on carrier
Processing Parameters	
DSI Cancellation	120 range, 5 Doppler bins
DSI Cancellation CPI	102400 samples (0.5s)
Range/Doppler Processing	120 range, 1601 Doppler bins
Range/Doppler CPI	819200 samples (4s)
CFAR Algorithm	GOCA-CFAR
CFAR Window	4 guard cells, 8 reference cells (either side of CUT)
CFAR Dimension	Doppler (Robust against bandwidth fluctuations)
CFAR Threshold	$P_{fa} = 10^{-5}$ (exponential noise model)

Due to licensing restrictions, a 10 W jammer could not be used within the FM band [150]. To overcome this, a very low power jamming signal was used (-35 dBm) and placed 24 m from the receive site as demonstrated in Figure 5.2. The jammer antenna used was a dipole antenna with a gain of 2.15 dBi. The placement of the jammer, 45 degrees between both the reference and surveillance antennas were chosen to ensure both the surveillance and reference channels experience the same jamming power levels. By adjusting the jammer transmit power, an effective JSR that allows comparisons between the measured and simulated results to be made.

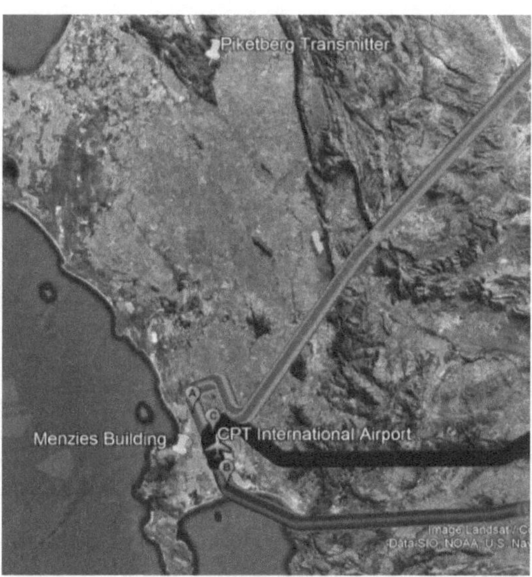

Figure 5.1: System geometry for field trials in the Western Cape of South Africa. The transmitter location is shown in the top center of the image labelled Piketberg Transmitter. The receiver site can be found in the center left of the map labelled Menzies Building. There are 3 flight paths that are most commonly used by commercial airliners, shown in green (A), red (B) and blue (C) as tracked by ADS-B data for the three targets detected in the trials.

Figure 5.2 shows the view of the radar receiver antennas from the jammer antenna location. It is noted that this form of ideal placement is highly unlikely due to the lack of knowledge of receiver location in real world scenarios. With this in mind, further experiments are required to evaluate the effects of jammer location relative to the receiver.

In order to compare the measured results with the simulated results shown in Section 3.6, the JSR on the reference and surveillance channels needs to be the same across both the simulated and measured scenarios. The direct signal power was calculated using (3.14) as shown in (5.1).

$$P_{r_{direct}} = \frac{P_t G_t G_r \lambda^2}{(4\pi R)^2} = \frac{(16\,400)(10^{7.2/10})(\frac{91.1\times10^6}{3\times10^8})^2}{(4\pi \times 118\,500)^2} \text{ W} \tag{5.1}$$
$$= -55 \text{ dBm}$$

The jammer power at the receiver was calculated in the same way using (5.1). With the jammer situated 45 degrees off boresight of the reference antenna, as well as being below it, the gain of the antenna was reduced by approximately 4 dB from 7.2 dBi to

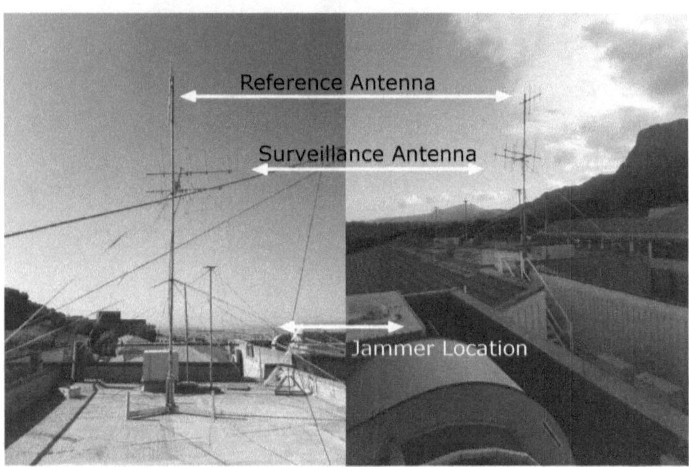

Figure 5.2: (Left) Reference and surveillance antennas set-up on the roof of the Menzies building at UCT. The top antenna was the reference antenna, pointing towards the Piketberg transmitter. The bottom two antennas were surveillance antennas, perpendicular to the reference antenna and pointing towards the Cape Town International airport. (Right) Radar antenna set-up as seen from the jammer location approximately 24 meters apart.

3.2 dBi. Transmitting at a power of -35 dBm results in a received power level of -89.6 dBm at the reference antenna before cable losses.

$$P_{r_{jam}} = \frac{(10^{-65/10})(10^{2.15/10})(10^{3.2/10})(\frac{91.1 \times 10^6}{3 \times 10^8})^2}{(4\pi \times 24)^2} \text{ W}$$

$$= -89.6 \text{ dBm}$$

(5.2)

This results in a Jamming-to-Signal ratio on reference channel ($JSR_{reference}$) of -31.6 dB with 2 dB of cable loss. This was directly comparable to the $JSR_{reference}$ in the simulated results for a 5 W jammer as shown in (3.18). The JSR_E varied from measurement to measurement depending on where the target was as well as targets incident angle. Assuming a calculated RCS of 23 dBsm, the resultant measured JSR_E was determined using (3.16). This gives a JSR_E of between approximately 40 dB and 46 dB depending on target location along the flight path and its incident angle. It is important to observe that this range of JSR_E was situation specific and will be different in reality because our simulation assumed a non-fluctuating target RCS of 23 dBsm. In our field-measurement results, there is, however, no means to accurately arrive at such an absolute target RCS value.

5.2 Measured Noise Jamming

The measurement was performed to compare the initial simulated results using Gaussian white noise jamming with measured data. The system was set-up as described in Section 5.1.

As demonstrated in Figures 5.3 and 5.4, a single target was visible in both ARD maps. Figure 5.3 shows a single Boeing 737-800 approaching the Cape Town international airport to land. Turning on the jammer and waiting 4 seconds (one CPI later), results in a raised noise floor as demonstrated in Figure 5.4 where the target echo was clearly masked slightly by the increased interference.

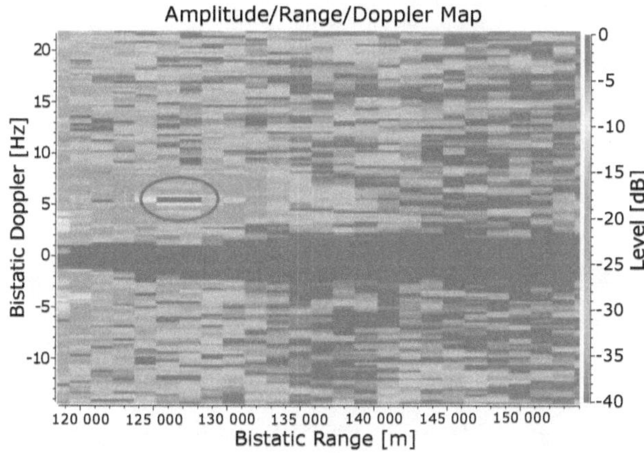

Figure 5.3: ARD map for measured system with no jamming applied. The target, a Boeing 737-800 flying towards the Cape Town international airport is shown in the red oval (127 km, 6 Hz). The targets' ADSB track is shown by the green track (A) in Figure 5.1 where $R_{Tx} = 112$ km and $R_{Rx} = 15$ km.

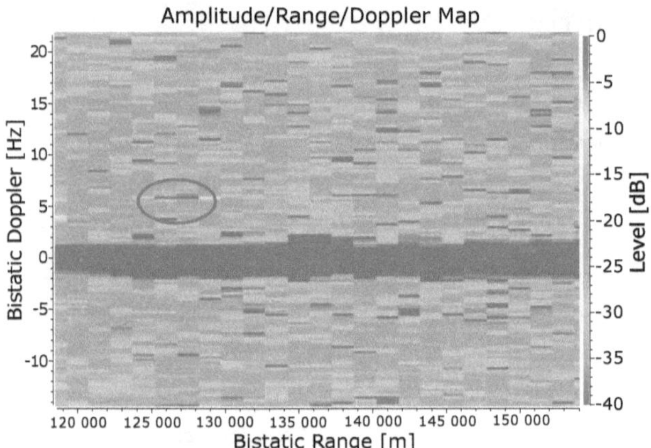

Figure 5.4: ARD map for measured system in the presence of AWGN jamming applied to both channels. The target is shown in the red oval and it is clear that there is a reduction in performance compared to the case of no jamming, demonstrated in Figure 5.3.

5.3 Measured Single Tone Jamming

Jamming of the PR through the use of a single tone on the carrier frequency was also investigated and compared to the simulated results. To achieve this, the system was set up such that the jammer transmits the waveform demonstrated in Figure 3.5.

With lack of funding to conduct a trial using cooperative targets flying a pre-defined flight path and an FM transmitter broadcasting the same content for each pass over, a compromise needed to be made. In order to maintain consistency in the results, each result needs to be evaluated on its own merits. To achieve this, the jammer was triggered on and off in 20 second bursts, equivalent to 5 CPIs. This allowed us to observe the performance of the system with a given target both with and without the presence of jamming. Obviously the target will move in the 20 second period and the broadcast content will be different, however it was felt that this at least allows the effects of the jamming to be observed on the same target.

Figure 5.5 demonstrates the output of the radar system with two targets clearly visible, as highlighted by the red ovals. Turning on the jammer and processing the data for another CPI results in Figure 5.6. The targets are highlighted by the red ovals where it was clear that the target was masked by the artefacts that appear across the entire range profile, as was the case in the simulated results in Figure 4.3.

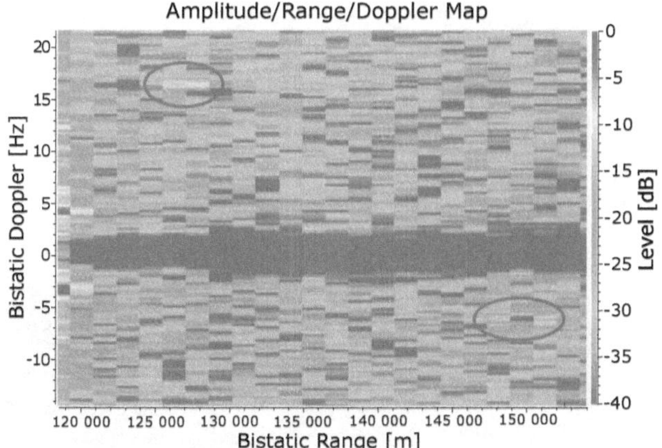

Figure 5.5: ARD map for measured system in the absence of jamming. Two targets are visible, highlighted by red ovals. The target at (-128 km, 15 Hz) is shown to be travelling towards the airport as it comes in to land, tracked by the blue track (C) in Figure 5.1, with $R_{Tx} = 121$ km and $R_{Rx} = 8$ km. The second target at (-149 km, -5 Hz) is shown to be travelling away from Cape Town by the red track (B) in Figure 5.1, with $R_{Tx} = 132$ km and $R_{Rx} = 17$ km.

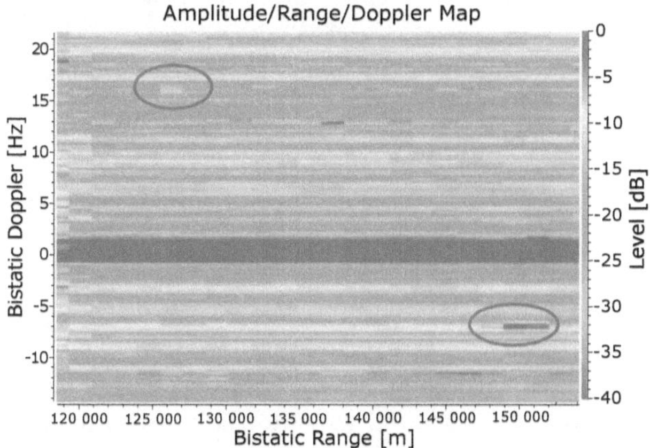

Figure 5.6: ARD map for measured system output in the presence of single tone jamming on the carrier frequency, applied to both the reference and surveillance channels. This map is produced 1 CPI (4 seconds) after the map demonstrated in Figure 5.5.

Passing the output ARDs through the CFAR filter in the Doppler dimension results in missed target detections. Figure 5.7 shows the result of applying the CFAR detector to the output ARDs of the measured system. The result was achieved by recording data in the absence of jamming for 20 seconds (5 CPIs) and then turning on the jammer, transmitting a single tone on the carrier for 20 seconds (5 CPIs). The jammer was then toggled off again for another 20 seconds. This was plotted in Figure 5.7 where the green arrows represent normal operation and the red arrows represent the time when the jammer was active. Both targets were clearly detected without jamming present and were completely masked when jamming was present.

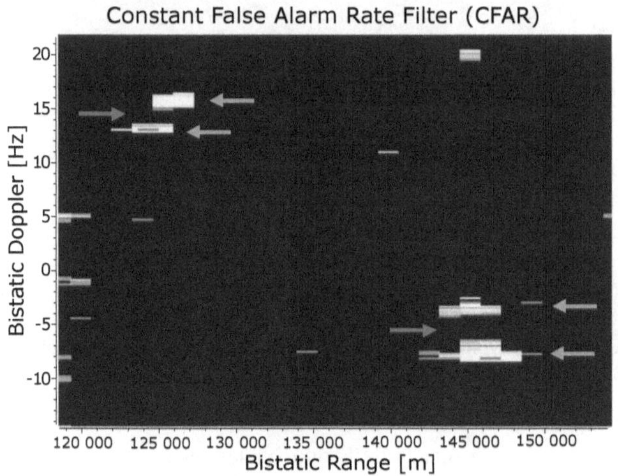

Figure 5.7: Combined CFAR output showing the two targets in Figures 5.5 and 5.6 as they progress along their flight paths. The green arrows indicate where the target is tracked in the absence of jamming. The red arrows indicate the missed detections when jamming is applied. The CFAR filter in this output is applied in the Doppler dimension.

5.4   Chapter Summary

To validate some of the simulated results shown in Chapter 4, a representative measurement was performed using a real FM PR. The PR used for the measurement campaign was a commercial PR from Peralex electronics, the ComRAD3 [113]. Due to lack of funding, a complete measurement campaign with cooperative targets could not be achieved, however measurements were taken to demonstrate the validity of the simulated results in Chapter 4.

To perform the measurement campaign, the real FM PR was placed on the rooftop of Menzies building at UCT. To circumvent having to transmit at relatively high power, the jammer was placed on an adjacent rooftop, 24 m from the PR receiver antennas. The jammer power was chosen such that the JSR between the reference transmitter and the jammer signal at the terminals of the PR receiver antenna was the same for both the measured and simulated results. Targets were then tracked for a period of time before the jammer was toggled on, this allowed for the targets to be clearly visible by the system to the targets being masked by the jamming. The jammer was then toggled off and the targets were once again detected by the system in their expected location as illustrated in Figure 5.7.

The measured results were shown to match the simulated results where the effectiveness of applying tone jamming to an FM PR was demonstrated. It was shown that peaks could be inserted across the range profile at various Dopplers when a tone was used as the jamming waveform, as was the case in the simulations. The system noise floor was also shown to increase as expected from the simulated scenario when noise jamming was applied to the FM PR.

Chapter 6

DVB-T2 Passive Radar

The second generation of terrestrial Digital Video Broadcast, called DVB-T2, is increasingly being deployed world-wide [82]. Germany, for example, is entirely serviced by DVB-T2. Over the past decade there have been numerous PR demonstrators designed and developed to operate with the original terrestrial digital video broadcast standard, DVB-T [65, 83–90]. Now, however, PR are being adapted to utilise the increasingly widespread DVB-T2 standard [91–95].

Digital broadcast protocols have the advantage that they have high and constant bandwidth compared to traditional analogue systems. The wider bandwidths provided by digital broadcast services yield finer PR range resolution, however these systems are not without their drawbacks as discussed in Section 6.1. DVB-T2 is an evolution of DVB-T that introduces a high level of flexibility to the standard. This higher level of flexibility allows for use of the standard in a wider range of transmission environments and support for higher data rate transmissions. Another advantage of digital transmissions such as DVB-T2 is that the reference signal can be 'demodulated' and 'remodulated' to create a perfect, noise free reference for use as a matched filter. The demodulation process is implemented using a similar process to what is found in a standard commercial TV sets. Once the signal has been demodulated, the process can be reversed and the signal can be remodulated to represent the exact signal that was transmitted but this time without any features of the propagation channel (noise and interference). The technique is commonly referred to as 'demod-remod'. The exact demod-remod process will not be covered in any further detail as it is not within the scope of this work however, it has been extensively covered in [93, 100, 103, 151] while the open standard can be found in [99].

6.1 DVB-T2 Signal Overview

The DVB-T2 signal is an OFDM signal with a pre-defined, standard structure where the number of sub-carriers is dependent on the operating mode. The DVB-T2 signal has 3 levels of abstraction, a super frame which carries multiple smaller T2 frames, each of which contain symbols carrying content data. The maximum length of a super frame is 63.75 seconds if Future Extension Frames (FEFs) are not used which corresponds to 255 T2 frames of 250 ms each. If FEFs are used, each super frame is 127.5 seconds long. The number of data symbols in each T2 frame depends on the length of each symbol. If, for example, each symbol is 32K (32 768 samples) long, then a total of 60 useful symbols can fit into each frame [99]. Each symbol carries data using a variation Quadrature Amplitude Modulation (QAM).

In order to work with a DVB-T2 signal, it is important understand how each T2 frame is constructed. The basic time domain structure of a T2 frame is demonstrated in Figure 6.1.

| P1 | G_{P2} | P2 | G_1 | Data Symbol 1 | G_2 | Data Symbol 2 | $\bullet\bullet\bullet$ | G_{n-1} | Data Symbol n-1 | G_n | Data Symbol n |

Figure 6.1: Single T2 frame structure illustrating the position of the P1 symbol, followed by the guard symbol and the P2 symbol before the data symbols.

The frame begins with the P1 symbol. The P1 symbol is used to speed up channel search and allows for basic timing and coarse frequency offset correction between the receiver and the transmitter. The P1 symbol is a 1K (1024 samples) OFDM symbol that consists of three distinct parts A, B, C. The P1 symbol also contains information concerning the FFT size (32K, 16K, 8K, 4K, 2K and 1K) of the remaining OFDM symbols (P2 and Data) within the frame and whether either MISO or SISO transmission is being used.

After the P1 symbol, a P2 symbol and a number of data symbols follow. The number of data symbols within the frame is dependent on the FFT size specified by P1. The P2 and data symbols contain pilot tones which are known reference values spaced at different frequency carriers and symbol indices within each symbol. Pilot tones are used by a receiver for channel estimation, timing and fine frequency offset correction.

The P2 symbols contain the relevant Layer-1 (L1) signalling, split into L1-pre signaling and L1-post signaling, as well as a highly dense pilot pattern. The number of pilot tones within a P2 symbol is much higher than the number of pilot tones in a data symbol and is dependent on the parameters given by P1. The L1 signalling provides the means to demodulate the DVB-T2 frame and extract information. This information includes the number of data symbols, the constellation size as well as the pilot pattern present within the data symbols.

The data symbols contain the data and pilots grouped into Physical Layer Pipes (PLP). The pilots can be further divided into Scattered Pilots (SP) and Continual Pilots (CP). The SP are spread in symbol and frequency bins in one of 8 pre-defined patterns using a Binary Phase Shift Keying (BPSK) mapping scheme (specified by the L1 information) as highlighted in Table C.1 in Appendix C. The CP are placed at the same sub-carrier index for all data symbols in a frame. The sub-carrier index of the CP is dependent on the scattered pilot pattern in use and the FFT size of each symbol. Each PLP in the data symbol contains their own constellation mapping (Quadrature Phase Shift Keying (QPSK), 16-QAM, 64-QAM or 256-QAM), where the number and configuration of each PLP is given by the L1 information. Any remaining cells in the DVB-T2 frame are filled with either auxiliary streams, which are transmitter specific and are not required for demodulation, or dummy cells, which are empty cells.

Table 6.1: DVB-T2 signal parameters used in the Cape Town Area

Parameters	Value
FFT Size	32768 E
Active Carriers	27841
Pilot Pattern	PP4
T_u	3584 μs
T_g	224 μs (1/16)
Subcarrier Spacing	279 Hz
Bandwidth	7.77 MHz
P2 Pilot Amplitude	$\sqrt{37}/5$
Continual Pilot Amplitude	8/3
Scattered Pilot amplitude	7/4
Scattered Pilot: Separation of pilot bearing carriers	12
Scattered Pilots: Number of symbols forming one scattered pilot sequence	2
P2 Encoding	64 QAM
Data Encoding	256 QAM
Number of Continual Pilots	178
Scattered Pilots per Symbol	1159+
Total Pilots per Symbol*	1196

* Where the scattered pilots fall on the same frequency bin as the continual pilots, the amplitude is set to the scattered pilot amplitude.

6.2 Range-Doppler Ambiguities

As with most OFDM based communication protocols, pilot carriers are utilised which have pre-defined patterns. These pilot carriers lead to ambiguities within the ARD map due to their periodicity and raised amplitude levels. To evaluate these ambiguities, the AF of a single DVB-T2 frame containing 32K symbols is illustrated in Figure 6.2. The AF is a two-dimensional function of time delay and Doppler shift and is described mathematically using (6.1):

$$\chi(\tau, v) = \int_{-\infty}^{\infty} r(t)r^*(t - \tau)e^{-j2\pi vt}dt \tag{6.1}$$

where τ and v represent the time delay and Doppler shift respectively. The peaks

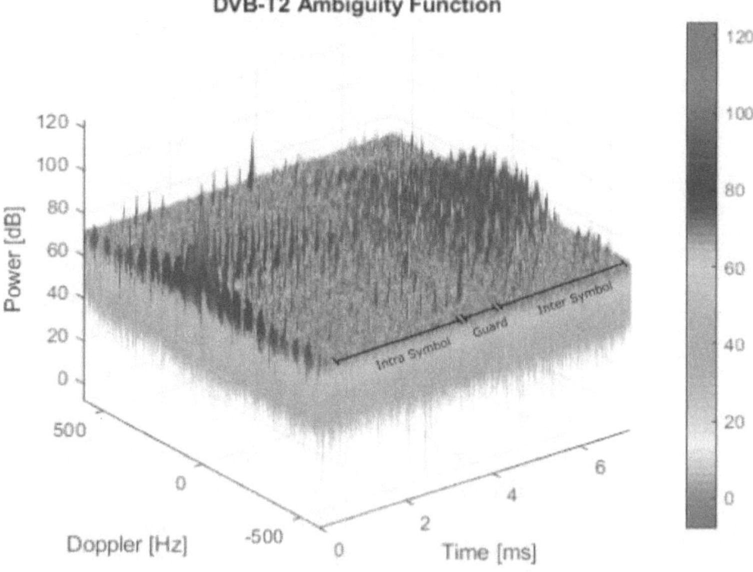

Figure 6.2: 3D AF of a frame within the DVB-T2 signal.

seen in Figure 6.2 are due to the boosted CP and SP within the frame (PP4 in this instance). The CP are on the same frequency bins throughout the DVB-T2 frame which creates low amplitude ridges across the range profile at fixed Doppler frequencies. While these ridges are not clearly visible above the noise floor in Figure 6.2, they are illustrated clearly in Figure 6.3 where the Cross-ambiguity Function (CAF) of the CP is shown. The SP can be seen scattered around the range-Doppler map in a pattern defined by the cyclic nature of their positioning within the DVB-T2 frame. The mathematical representation for the positions of the ambiguities is described in detail in [83, 84, 86, 89]. The ambiguities and their influence on target detection depends heavily on the transmit mode and pilot pattern used. The three main ambiguity regions are highlighted as intra-symbol, guard and inter-symbol ambiguities as demonstrated in Figure 6.2.

6.2.1   A. Intra-symbol Ambiguities

Intra-symbol ambiguities refer to ambiguities corresponding to $0 < \tau < T_u$. This leads to ambiguities in both range and Doppler. These peaks are due to the boosted pilot amplitudes relative to the data symbols as highlighted in Table C.1 in Appendix C. The intra-symbol ambiguities need to be addressed with regardless of the symbol length as they are always present in the system unlike the other ambiguities which may or may not appear depending on the symbol length.

Range Ambiguities
The SP pattern causes peaks in fast-time that repeat every $\frac{1}{P_s \cdot C_s}$ seconds, where P_s is the number of carriers between each pilot and C_s is the carrier spacing.

Doppler Ambiguities
Since the CP pattern repeats every N symbols, there is also periodicity over slow-time. The symbol length, T_u, corresponds to a periodicity every $\frac{1}{N \cdot T_u}$ Hz in Doppler.

6.2.2   B. Guard Interval Ambiguities

Guard interval ambiguities are a result of the guard intervals at delays equal to T_u. The guard intervals between symbols are repetitions of the end of each succeeding symbol. These repetitive symbols result in ambiguities that only appear at regular intervals equal to T_u as seen in Figure 6.2 where $\tau = 3.584$ ms for a 32K symbol with a 2K guard [93]. In this case, the guard interval ambiguities appear well outside the detection range of the system at greater than 1 000 km.

6.2.3   C. Inter-symbol Ambiguities

Inter-symbol ambiguities refer to ambiguities corresponding to $\tau > T_u$. Inter-symbol ambiguities arise from SP pattern repetition every N symbols, which results in additional ambiguities appearing at a delays of $n \cdot N \cdot T_u$ seconds. For systems utilising 1K,

2K or 4K modes, this potentially falls within the detection range however, when 8K, 16K or 32K modes are used, these ambiguities fall well outside the detection range of a typical system as highlighted in Table 6.2 where the guard interval range represents the bistatic range at which guard interval and inter-symbol ambiguities will appear.

Table 6.2: DVB-T2 OFDM symbol guard intervals for different FFT sizes.

FFT Size	T_u [ms]	Guard Interval Range [km]	Guard Interval [ms]						
			1/128	1/32	1/16	19/256	1/8	19/128	1/4
1k	0.112	34	NA	NA	0.007	NA	0.014	NA	0.028
2k	0.224	67	NA	0.007	0.014	NA	0.028	NA	0.056
4k	0.448	134	NA	0.014	0.028	NA	0.056	NA	0.112
8k	0.896	268	0.007	0.028	0.056	0.0665	0.112	0.133	0.224
16k	1.792	537	0.014	0.056	0.112	0.133	0.224	0.266	0.448
32k	3.584	1075	0.028	0.112	0.224	0.266	0.448	0.532	NA

6.3 Quantifying Pilot Effects

Since the AF is a linear process, it can be broken down into the individual contributions of each component. To get a comprehensive understanding of the individual effects of each component within the DVB-T2 signal, the AF of each component including the data symbols, the CP, the SP and the P2 symbol pilots is calculated.

6.3.1 Continual Pilot Effects

Figure 6.3 illustrates the CAF of the CP after demod-remod where all other signal components set to zero and the original demod-remod signal. The CP pattern consists of raised carriers appearing on the same carriers across time. This causes ridges that appear along delay. Where a SP falls on the same carrier as the CP, the CP is set to the same amplitude as the SP. This results in a pulse train along the carrier where the two pilots combine, leading to an additional peak forming on top of the CP ridge in the same arrangement as the SP peaks in Figure 6.4. The contribution of the CP ridges to the AF is relatively small since they appear only 2 dB above the noise floor. The peaks caused by the SP appearing on the same carriers as the CP, however, appear 13 dB above the noise floor. The peak levels are governed by the amplitude level of the pilots which change depending on the FFT size as highlighted in Table C.2 in Appendix C.

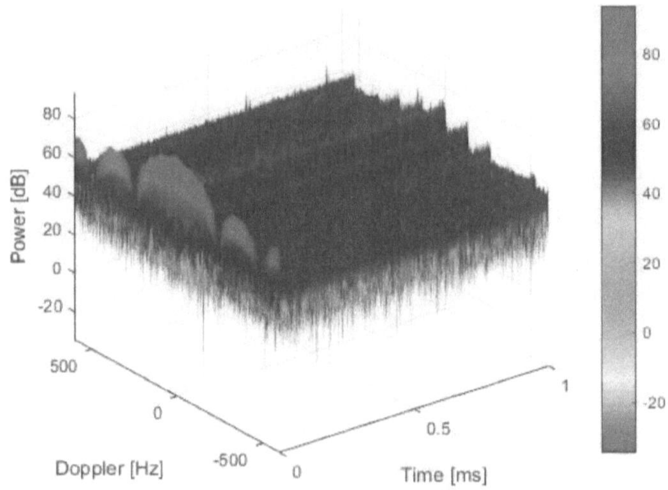

Figure 6.3: Continual pilot ambiguity function

6.3.2 Scattered Pilot Effects

Figure 6.4 illustrates the CAF of the SP where all other carriers set to zero with the original demod-remod signal.

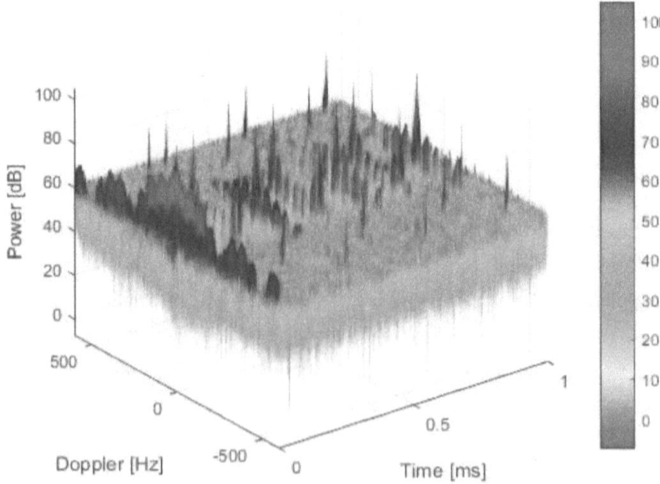

Figure 6.4: Scattered pilot ambiguity function

As with the AF demonstrated in Figure 6.2, the SP AF exhibits peaks which are periodic in both range and Doppler. The SP contribute significant power to the AF as the peaks appear 33 dB above the noise floor, compared to 13 dB for the CP.

Along with the guard interval ambiguities (which are not shown here due to their delay being impractical for a real 32K system), the final contribution to the AF is the P2 pilot pattern.

6.3.3 P2 Pilot Effects

Figure 6.5 illustrates the CAF of the P2 symbol pilots after demod-remod where all other components set to zero with the original demod-remod signal.

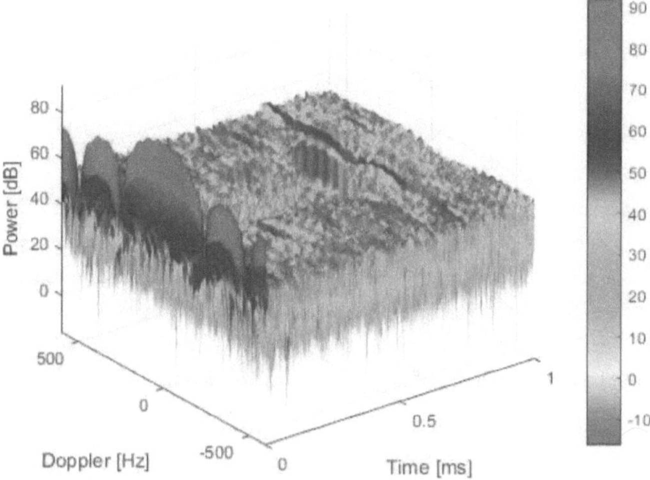

Figure 6.5: P2 pilot ambiguity function.

It is clear from Figure 6.5 that the P2 symbol contributes almost nothing to the complete AF and can therefore be ignored for the purposes of this work.

6.4 Typical DVB-T2 Passive Radar Processing Chain

The processing chain of a DVB-T2 PR is significantly more complex than that of an FM PR. This added complexity is due to the OFDM waveform in use which leads to ambiguities in the ARD map that need to be removed. As shown in Section 6.1, these ambiguities appear at different locations depending on the pilot pattern used by the transmitter. Since the pilot signals are completely deterministic, their exact locations can be determined through the demodulation process where they can subsequently be removed through processing.

Using OFDM based DVB-T and DVB-T2 transmissions as a PR illuminator of opportunity has been investigated by numerous researchers, most notably [83–86, 89, 90]. There are two common approaches to processing OFDM based signals in PR, the first of which is known as mismatched filtering, proposed in [83, 84] and further explored in [86, 89]. Mismatched filtering is a traditional cross-convolution based approach which involves demodulating the reference signal and then remodulating it to obtain a noise-free, slightly modified reference. This new reference signal is then used as a matched filter to perform range-Doppler processing. Since the new reference signal is a modified version of the original, it is no longer considered matched to the original and therefore the term 'mismatched filtering' is used.

The second approach, shown in [85] and expanded on in [90], is referred to as inverse filtering. Inverse filtering is a process whereby the signal undergoes demod-remod to produce an exact noise-free copy of the reference signal. This noise-free reference signal is then used to perform the range-Doppler processing by first performing a carrier-wise division of the OFDM symbols in the surveillance channel using the same OFDM symbols in the remodulated reference channel. A 2D FFT is then applied to produce an ARD map. This in effect normalises the carriers and compresses all the direct signal clutter into the zero-Doppler bin of the ARD.

Figure 6.6 illustrates the core processing steps required in a DVB-T2 PR. The black blocks in Figure 6.6 are common to both mismatched and inverse filtering. The red blocks are specific to mismatched filtering while the blue blocks are specific to inverse filtering. The exact location within the processing chain of the DSI cancellation blocks is discussed further in Section 6.4.3.

6.4.1 Mismatched Filtering

As mentioned Section 6.4, Mismatched filtering is a classic CAF approach whereby the reference signal is cross-correlated with the surveillance signal according to (6.2).

$$\chi_{rs}(\tau, v) = \int_{-\infty}^{\infty} r(t)^* s(t + \tau) e^{-j2\pi v t} dt \tag{6.2}$$

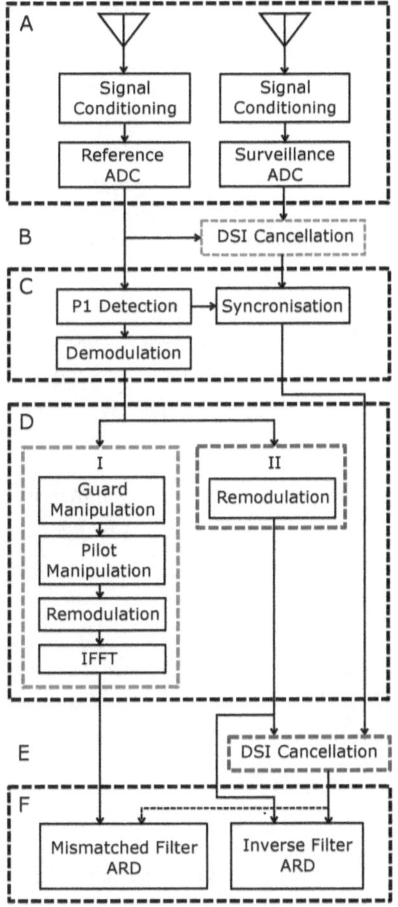

Figure 6.6: DVB-T2 PR processing block diagram.

This is typical of how a classic FM PR processing chain operates where the range and Doppler resolutions are related to the bandwidth of the signal and the integration time respectively. This leads to a range resolution of approximately 40 m and a single frame Doppler resolution of approximately 4 Hz. Mismatched filtering has been demonstrated in [83, 84, 86, 89] to be an effective means of processing OFDM based DVB-T signals. Mismatched filtering differentiates itself from matched filtering in that the reference signal used in the range-Doppler processing step is a modified version of the original rather than an exact copy. This modified or 'mismatched' reference signal allows the removal of the signal ambiguities within the ARD map. It has been demonstrated that mismatched filtering offers a way to effectively mitigate the effects of pilot ambiguities in the ARD map to enhance target detections. The major drawback however, is that it is computationally expensive due to the convolution step in (6.2). This can be mitigated slightly by using an FX batches processor (typically suitable for higher bandwidth signals) whereby the FFT is taken from the output of the demod-remod stage and then the cross-correlation process is performed as discussed in [68].

Ambiguity Removal with Mismatched Filtering

As described in Section 6.1, the DVB-T2 signal has ambiguities that need to be removed in order to successfully detect and track targets. These pilot ambiguities can be removed during the CAF process as shown in [93]. To achieve this, the pilot signal levels need to be manipulated in the remodulation process to form the mismatched reference signal.

As shown in Section 6.1, there are three so-called ambiguity regions which arise as a result of the signal structure. Gao [84] and Harms [86] describe the processes required to remove the different ambiguities in each region.

A. Intra-symbol Ambiguity Removal

It is mentioned in [93] and demonstrated in [92] that these ambiguities can be removed by blanking the pilot carriers, however, since the ambiguities in the intra-symbol region are due to the relative amplitude differences between the pilot and the data carriers, as highlighted in Table C.1 in Appendix C, they are therefore removed through a normalisation process. This normalisation is achieved by modifying the pilot levels in the remodulating phase to $1/A_{\mathrm{SP}}$. This then normalises the pilots to the background carrier levels during the correlation process, resulting in the removal of the intra-symbol pilot ambiguities. Normalising the pilots in such a way results in a slight (approximately 0.44 dB) decrease in integration gain (in a 32K system) as shown in (6.3).

$$
\begin{aligned}
G_{loss} &= 10 \cdot \log_{10}\left(\frac{\text{Active Carriers} - \text{Pilot Carriers}}{\text{Active Carriers}}\right) \\
&= 10 \cdot \log_{10}\left(\frac{27841 - 1196}{27841}\right) \\
&= -0.44 \text{ dB}
\end{aligned}
\tag{6.3}
$$

Setting the pilot amplitudes to zero, as suggested in [92] and [93], can be modeled as an inverted pulse train, leading to additional ambiguities. As these ambiguities are originally a result of the amplitude differences between the pilot and data carriers, setting them to zero once again results in ambiguities within the ARD map as described by (6.6).

Assuming the spectrum of a single symbol has no zero-value carriers, i.e. contains only data carriers and pilots, the blanking of these pilots can be modeled by multiplying the spectrum by a pulse train of zeros and ones which follows the SP pattern as demonstrated in Figure 6.7.

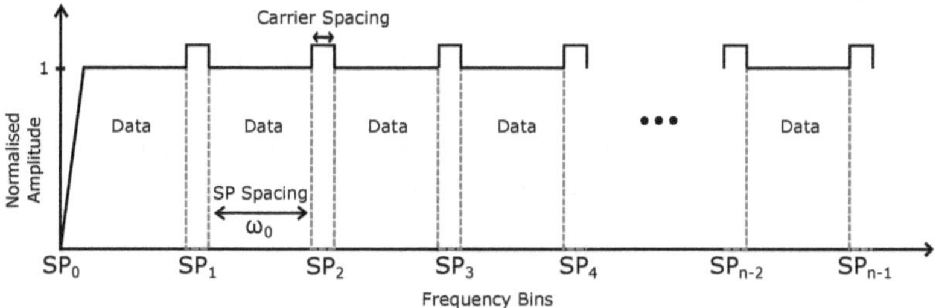

Figure 6.7: OFDM signal structure in frequency domain.

Blanking the carriers at regular time intervals corresponding to the SP pattern is the same as multiplying the spectrum by a pulse train in the frequency domain. This is shown in (6.4) where the pulse period is $\frac{1}{(P_s)(C_s)}$ with P_s representing the pilot spacing and C_s representing the carrier spacing.

$$S_{out}(f) = S_{data}(f) \times S_{blank}(f) \tag{6.4}$$

where $S_{blank}(f)$ is described as:

$$
\begin{aligned}
S_{blank}(f) &= 1 - \omega_0 \sum_{n=0}^{N_{SP}-1} \delta(\omega - n\omega_0) \\
&= 1 - 2\pi(P_s)(C_s) \sum_{n=0}^{N_{SP}-1} \delta(2\pi f_0 - n2\pi(P_s)(C_s))
\end{aligned}
\tag{6.5}
$$

$$S_{blank}(t) = \delta(t) - \underbrace{\sum_{n=0}^{N_{SP}-1} \delta\left(t - n \cdot \frac{1}{(P_s)(C_s)}\right)}_{\text{Pulse train with period } \frac{1}{(P_s)(C_s)}} \tag{6.6}$$

The pulse period, T, which defines where the peaks will appear is therefore:

$$T = \frac{1}{(P_s)(C_s)} \tag{6.7}$$

The ambiguities arising from blanking the pilots will therefore be located in the same position in the ARD map as the original boosted pilot ambiguities.

B. Guard Interval Ambiguity Removal

Unlike the intra-symbol ambiguities that arise due to the relative level differences between the pilot and data carriers, the guard interval ambiguities arise due to the cyclic repetition of the guard symbols. As a result, these ambiguities can be removed by blanking the guard intervals in the remodulation step as demonstrated in [84, 86, 89]. This however, comes with a loss of integration gain relative to the length of the guard interval itself. Table 6.2 highlights the bistatic ranges at which the guard interval ambiguities are present for a given symbol length.

C. Inter-symbol Ambiguity Removal

As with the ambiguities caused by the guard intervals, ambiguities appearing in the inter-symbol region are a result of their repetition over different symbols and as such, can be removed by blanking the pilots. In most scenarios, as with the guard intervals, the inter-symbol ambiguities will fall well outside the instrumented range of the radar and can therefore be ignored.

Since the process of blanking the pilots and normalising the pilots are counteractive, two parallel processes need to be initialised whereby one normalises and the other blanks the pilots. Each ambiguity region then needs to be calculated in parallel using the appropriately modified reference function with the two results stitched together to form an ambiguity-free ARD map as demonstrated in [84].

Processing Gain

As with FM PR, the processing gain is a function of the signal bandwidth and the CPI length. The processing gain for a DVB-T2 PR is therefore given as

$$G_p = B \cdot T \tag{6.8}$$

where B is the occupied bandwidth of the signal and T is the integration time. For a DVB-T2 signal we can determine the integration gain based on the number of symbols used in the creation of each ARD map. If we were to use a single frame with a 32K symbol length, the system would achieve and integration gain of:

$$G_p = 10 \cdot \log_{10}(N_{active} \cdot C_s \cdot N_{symbols} \cdot T_s) - G_{loss}$$
$$= 10 \cdot \log_{10}(27\,841 \cdot 279\text{Hz} \cdot 60 \cdot (3.584 + 0.224)\text{ ms}) - 0.44\text{ dB} \qquad (6.9)$$
$$= 62\text{ dB}$$

where N_{active} is the number of active carriers, C_s is the carrier spacing, $N_{symbols}$ is the number of symbols used per ARD map and T_s is the time length of each symbol including the guard interval if used. G_{loss} is the loss in integration gain due to the normalisation of pilots as shown in (6.3).

6.4.2 Inverse Filtering

Inverse filtering is a computationally efficient technique for producing ARD maps that exploits the OFDM nature of the transmitted signal. This is accomplished by 'normalising' the surveillance channel relative to the reference channel. To demonstrate this, consider the simplified representation of the direct signal at the receiver shown in (6.10):

$$s_{ds}(t) = e^{i2\pi f_0 t} \sum_{n=0}^{N-1} \left[A_n e^{i\Phi_n} e^{i2\pi(-\frac{BW}{2} + n\frac{BW}{N-1})t} \right] \qquad (6.10)$$

where BW represents the total bandwidth occupied by active carriers, N is the number of active carriers in use, f_0 is the carrier frequency, and $A_n e^{i\Phi_n}$ makes up the complex amplitude for each carrier, n. Windowing and symbol numbering have been omitted for clarity. Echoes from potential target will be time delayed copies of (6.10), seen here in (6.11):

$$s_{echo}(t) = s_{ds}(t - t_d)$$
$$= e^{i2\pi f_0(t-t_d)} \sum_{n=0}^{N-1} \left[A_n e^{i\Phi_n} e^{i2\pi(-\frac{BW}{2} + n\frac{BW}{N-1})(t-t_d)} \right] \qquad (6.11)$$

with

$$t_d = \frac{R_0}{c} + \frac{v_{bis}}{c}t \qquad (6.12)$$

where R_0 is the difference in bistatic distance at the start of the symbol, v_{bis} is the bistatic velocity for this target, and c is the propagation speed. After down conversion, the baseband representations of the direct and echo signal becomes:

$$s_{ds,bb} = \sum_{n=0}^{N-1} \left[A_n e^{i\Phi_n} e^{i2\pi(-\frac{BW}{2} + n\frac{BW}{N-1})t} \right] \qquad (6.13)$$

and

$$s_{echo,bb} = e^{-i2\pi f_0 t_d} \sum_{n=0}^{N-1} \left[A_n e^{i\Phi_n} e^{i2\pi(-\frac{BW}{2}+n\frac{BW}{N-1})(t-t_d)} \right]$$

$$= e^{-i2\pi f_0 \frac{R_0}{c}} e^{-i2\pi f_0 \frac{v_{bis}}{c} t}. \tag{6.14}$$

$$\sum_{n=0}^{N-1} \left[A_n e^{i\Phi_n} e^{i2\pi(-\frac{BW}{2}+n\frac{BW}{N-1})(1-\frac{v_{bis}}{c})t} \cdot e^{-i2\pi(-\frac{BW}{2}+n\frac{BW}{N-1})\frac{R_0}{c}} \right]$$

The time-changing delay term, t_d, has three notable effects. It induces a Doppler shift $\left(e^{-i2\pi f_0 \frac{v_{bis}}{c} t}\right)$, an additional phase term that is determined by the bistatic distance $\left(e^{-i2\pi f_0 \frac{R_0}{c}}\right)$, and a shift in phase of each individual sub-carrier $\left(e^{-i2\pi(-\frac{BW}{2}+n\frac{BW}{N-1})\frac{R_0}{c}}\right)$. The sub-carrier phase shift, which is directly proportional to the bistatic distance of the target, is what inverse filtering relies on for range processing.

The normalisation of the echo signal is where the term inverse filtering or reciprocal filtering is derived. This normalisation is applied in the frequency domain, where the complex amplitudes of the echo signal spectrum are divided, element wise, by the complex amplitudes of the direct signal spectrum as shown in (6.17).

$$S_{ds,bb}(k) = A_k e^{i\Phi_k} \tag{6.15}$$

$$S_{echo,bb}(k) = A_k e^{i\Phi_k} \cdot e^{-i2\pi f_0 \frac{R_0}{c}} \cdot e^{-i2\pi(-\frac{BW}{2}+k\frac{BW}{N-1})\frac{R_0}{c}} \cdot \text{sinc}\left(\frac{\pi f_0 N}{BW} \cdot \frac{v_{bis}}{c}\right) \tag{6.16}$$

$$S_{norm,bb}(k) \triangleq \frac{S_{echo,bb}(k)}{S_{ds,bb}(k)}$$
$$= e^{-i2\pi f_0 \frac{R_0}{c}} \cdot \text{sinc}\left(\frac{\pi f_0 N}{BW} \cdot \frac{v_{bis}}{c}\right) \cdot e^{-i2\pi(-\frac{BW}{2}+k\frac{BW}{N-1})\frac{R_0}{c}} \tag{6.17}$$

The resultant output of the inverse filter is a rotating phasor where the speed of rotation is directly proportional to the bistatic distance of the target echo. Inter-carrier interference (caused by the shift in frequency of the echo signal) will be neglected for the time being. Range processing is accomplished by taking the Discreet Fourier Transform (DFT) of $S_{norm,bb}$:

$$S_{range}(p) \triangleq DFT\{S_{norm,bb}\} \tag{6.18}$$

The maximum amplitude of S_{range} occurs at:

$$p_{max} \approx BW \frac{R_0}{c} \tag{6.19}$$

Successively processing the adjacent OFDM symbols yields a result that is similar to the prior symbol, differing only in phase due to the change in R_0:

$$S_{range,\,symbol\,(y+1)} \approx S_{range,\,symbol\,(y)} \cdot e^{-i2\pi f_0 \frac{R_\Delta}{c}} \tag{6.20}$$

The change in bistatic distance is denoted here as R_Δ. Packing subsequent range lines into a rows within a matrix therefore forms an additional rotating phasor down each column:

$$\begin{aligned}
s_{doppler}(p, y) &\triangleq S_{range,\,symbol\,(y)}(p) \\
&= S_{range}(p) \cdot e^{-i2\pi f_0 \frac{R_\Delta}{c} y}
\end{aligned} \tag{6.21}$$

The speed of rotation will, in this case, be determined by the phase term $e^{-i2\pi f_0 \frac{R_\Delta}{c}}$. Taking a DFT in the slow-time dimension (down the matrix columns) yields the ARD map:

$$ARD(p, q) = DFT\big\{s_{doppler}(p, y)\big\} \tag{6.22}$$

The maximum amplitude of $ARD(p, q)$ occurs at:

$$q_{max} = \frac{f_0 R_\Delta N_{sym}}{c} \tag{6.23}$$

where N_{sym} is the number of consecutive OFDM symbols processed for each ARD map.

Limitations of Inverse Filtering

As the process of inverse filtering relies on the OFDM nature of the signal, it is highly dependant on the exact specifications used within the transmission, specifically the FFT length of each symbol and the resulting guard symbol length which directly limits the maximum unambiguous range and Doppler of the system.

This means that for short OFDM symbols such as 1K, the maximum unambiguous range is severely limited when compared to longer symbols such as 16K or 32K as highlighted in Table 6.2. The upside to this is that the maximum unambiguous Doppler of the system is limited by the inverse of the symbol length. This means that for short symbol lengths (1K, 2K, 4K or even 8K), the unambiguous Doppler is large and therefore does not pose any issues for typical applications. For larger symbol lengths such as 16K and 32K however, the maximum unambiguous Doppler is severely limited.

Processing Gain

As with mismatched filtering, the integration gain can be determined by the number of symbols used in the creation of each ARD map. If a single frame with a 32K symbol length were to be used, the system would achieve and integration gain of:

$$\begin{aligned}
G_p &= 10 \cdot \log_{10}(N_{active} \cdot C_s \cdot N_{symbols} \cdot T_u) \\
&= 10 \cdot \log_{10}(27\,841 \cdot 279 \text{ Hz} \cdot 60 \cdot 3.584 \text{ ms}) \\
&= 62.2 \text{ dB}
\end{aligned} \tag{6.24}$$

where N_{active} is the number of active carriers, C_s is the carrier spacing, $N_{symbols}$ is the number of symbols used per ARD map creation and T_s is the time length of each symbol including the guard interval.

6.4.3 DSI Cancellation in DVB-T2 Passive Radar

Depending on where the DSI cancellation is applied, the results could vary slightly. If, as is the case with FM PR, the DSI cancellation is applied using the raw received signal, part of the jammer signal could be removed and performance could potentially be improved as demonstrated in Chapter 3 [152].

While the exact location of the DSI cancellation step within the processing chain can vary, if it were to be implemented using the raw data, a CGLS-like canceller would be needed since more advanced techniques such as ECA-CD require demodulation to be used [75]. To ensure real-time operation, a CGLS canceller would need to process the data in such a way that once convergence is achieved, it only iterates once or twice per symbol. This approach will allow for fast processing times while also ensuring effective cancellation due to the slow varying nature of the DVB-T2 signal. The disadvantage to this approach would be that the canceller would not be able to adapt to a rapidly changing clutter environment. The canceller would also not be able to adapt if the signal itself changes from frame to frame as is allowed for in a DVB-T2 transmission [99].

If the DSI canceller were to be applied after the remodulation stage, all of the jammer signal would be removed from the reference channel, resulting in the canceller having no additional effect on the performance of the system as shown in Chapter 3. If the system were to utilise a single antenna for both reference and surveillance channels, whereby the reference signal is extracted through the demod-remod process, the canceller would only be capable of suppressing the reference DSI and would not be capable of removing any additional interfering signals.

Since DSI cancellation techniques such as ECA-CD rely on successful demodulation of the reference signal, if the jammer were capable of disrupting the reference signal to the point where demodulation is no longer possible, the entire remodulation process would fail. In the case of remodulation failure, inverse filtering would be rendered inapplicable and the mismatched filtering approach would be forced to revert to standard matched filtering. This type of attack could be achieved by jamming the P1 symbol, however, this would require significant power as is discussed in Chapter 7.

It is therefore feasible to assume that most DVB-T2 systems will either not use a DSI canceller (as is currently the case due to the low zero-Doppler sidelobe levels within the DVB-T2 AF) or would use a highly efficient canceller such as ECA-CD which requires the demod-remod process in order to work. As a result, any DSI canceller would not, if used in such a way, have any additional impact on system performance in the presence of jamming.

6.5   Chapter Summary

This chapter has provided a brief overview of the DVB-T2 signal in the context of PR. The contributions of the various pilot signals within the DVB-T2 signal have been discussed and their effects on the resultant AF demonstrated. It has been demonstrated that the most significant contribution to the ambiguities within the AF are the scattered pilots. While the continual pilot pattern produces ridges at constant Doppler at particular ranges, their contribution to the AF is relatively small due to there being so few continual pilot carriers per OFDM symbol.

The three ambiguity zones namely: intra-symbol, guard interval and inter-symbol ambiguities have been demonstrated. The process of removing the ambiguities from each zone during the creating of the ARD map has been demonstrated using both mismatch filtering and inverse filtering techniques. As a result, a comprehensive review of these two processing techniques has been presented where the advantages and disadvantages of each technique has been demonstrated.

A brief discussion on DSI cancellation techniques in DVB-T2 PR has been presented, outlining the advantages and disadvantages of time-domain and frequency-domain based approaches.

Chapter 7

Electronic Attacks on DVB-T2 Passive Radar

Since passive radars radiate no EM energy, they are inherently difficult to detect using conventional ES techniques. The DVB-T2 signal is designed to operate in single frequency networks and can therefore handle high powered interference in such a way that demodulation can still take place. If an attacker were to jam the DVB-T2 signal in such a way to prevent demodulation, the P1 symbol would need to be attacked. In general, it has to be assumed that an attacker does not know the exact whereabouts of the PR receiver(s). As a result, it is difficult to inject a signal of sufficient power into the reference channel such that demodulation is no longer possible. For these results, it is therefore assumed that the jamming signal appears only in the surveillance channel since the jamming signal is removed through the demod-remod stage.

As with FM PR, the most obvious approach to attacking a DVB-T2 PR is through the use of noise jamming. Noise jamming has been demonstrated to be a relatively effective means of attacking an FM PR. It has been demonstrated that raising the noise floor of particularly the surveillance channel with increased noise jamming power is critical when trying to mask a target.

As the processing gain of a DVB-T2 PR is greater than 60 dB, in order to mask a target, significant levels of noise jamming would be needed across a relatively wide bandwidth. A more practical approach would be to attack the deterministic components of the DVB-T2 signal which could potentially lead to significant effects within the ARD map while using relatively little power.

As shown in [134] and [135], OFDM based DAB and DVB-T PR can be effectively attacked using the deterministic components of each signal. A similar approach is therefore followed in demonstrating 5 attacks for DVB-T2 PR using both mismatched filtering and inverse filtering. These attacks include:

- Noise Jamming

- Full Pilot Attack

- Continual Pilot Attack

- Scattered Pilot Attack

- Pulsed Jamming

Unless otherwise stated, the results presented were achieved using real recorded DVB-T2 signals with the parameters highlighted in Table 6.1. The recorded signals were from a single channel receiver and therefore the second surveillance channel is simulated with three targets inserted. The simulated target parameters are highlighted in Table 7. The ARD map is calculated to ±200 Hz and 12.6 to 100 km as this falls within the capabilities of the system being developed at UCT as described in [94].

Table 7.1: Simulated target parameters.

Target	Bistatic Range [km]	Bistatic Doppler [Hz]	SNR [dB]
1	40	-35	-40
2	25	20	-35
3	80	140	-30

7.1 Jammer-to-Signal Ratio

As with the FM PR, one of the challenges in quantifying the performance of the DVB-T2 PR in the presence of jamming is defining the manner in which the JSR is presented. To define the JSR in the DVB-T2 context, we borrow from [134] and [135] where the authors define the JSR as the ratio of the jammer power to the total signal power in the surveillance channel. To elaborate on this, if the JSR is set at 0 dB, this implies that the average jammer power within the surveillance channel is equal to the average DVB-T2 signal power within the surveillance channel for each CPI.

This approach is taken partly to allow for comparison between the results in [134] and [135] as well as allowing the results to be referred back to the target echo power levels as was the case in Chapter 3.

7.2   Electronic Attacks using Mismatched Filtering

To determine a performance baseline, Figure 7.1 illustrates the resultant ARD map containing the three targets simulated using the parameters summarised in Table 7. All three targets were clearly visible at their expected locations as highlighted by the red ovals. As the surveillance channel is simulated, there is no zero-Doppler clutter other than the DSI in the zero-range bin.

The three targets sit -40 dB, -35 dB and -30 dB below the peak signal as expected while the noise floor is at approximately -60 dB. This means that the three targets sit 20 dB, 25 dB and 30 dB above the noise floor respectively when no DSI cancellation is applied.

Table 7.2: Simulated jammer parameters for Figure 7.1.

Attack	Range Shift [km]	Doppler Shift [Hz]	JSR [dB]
None	-	-	-

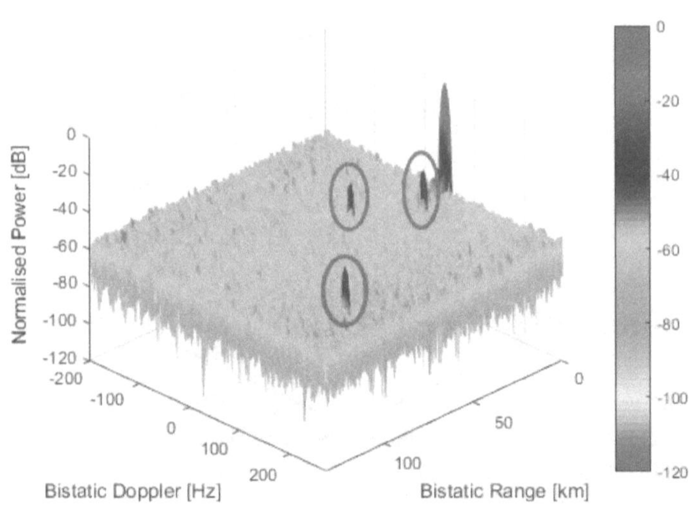

Figure 7.1: Reference plot for mismatched filtering.

The normalisation of the pilot signals achieved through mismatched filtering also reduces the level of any potential pilot attack signals. This normalisation process reduces the pilot levels by an amount depending on each pilots' relative boost. For the signal parameters in Table 6.1, the relative CP and SP power reductions are:

CP power reduction:

$$\begin{aligned}
\text{CP reduction} &= 20 \cdot \log_{10}\left(\frac{1}{\text{Boosted CP level}}\right) \\
&= 20 \cdot \log_{10}\left(\frac{3}{8}\right) \\
&= -8.52 \text{ dB}
\end{aligned} \tag{7.1}$$

SP power reduction:

$$\begin{aligned}
\text{SP reduction} &= 20 \cdot \log_{10}\left(\frac{1}{\text{Boosted SP level}}\right) \\
&= 20 \cdot \log_{10}\left(\frac{4}{7}\right) \\
&= -4.86 \text{ dB}
\end{aligned} \tag{7.2}$$

7.2.1 Mismatched Filtering - Noise Jamming

Noise jamming is applied to the surveillance channel with the JSR $= 0$ dB. This implies that the average noise power is equal to the average DVB-T2 signal power (direct signal plus simulated target echos) in the surveillance channel. This results in a slight increase in the noise floor as demonstrated in Figure 7.2.

Table 7.3: Simulated jammer parameters for Figure 7.2

Attack	Range Shift [km]	Doppler Shift [Hz]	JSR [dB]
AGWN	-	-	0

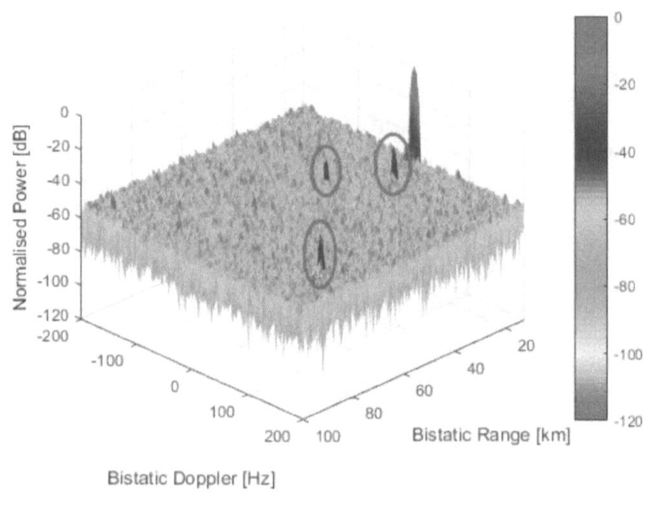

Figure 7.2: Mismatched filtering with noise jamming at 0 dB JSR.

Since adding noise to the system is a linear operation, to mask the targets the noise power needs to be at a JSR equal to the target levels above the system noise floor. In this case, the targets will be masked when the JSR of the noise jammer approaches 20 dB as demonstrated in Figure 7.3 where the JSR for the noise jammer is 10 dB and the weaker target can barely be seen above the raised noise floor.

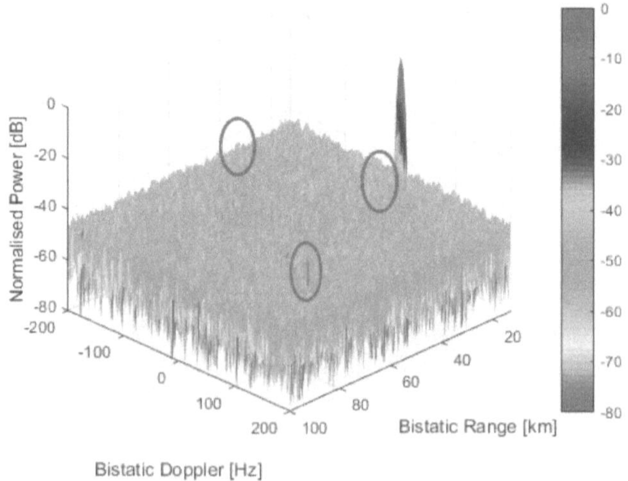

Figure 7.3: Mismatched filtering with noise jamming at 10 dB JSR.

Table 7.4: Simulated jammer parameters for Figure 7.3.

Attack	Range Shift [km]	Doppler Shift [Hz]	JSR [dB]
AGWN	-	-	10

The broadband and continuous nature of the noise jammer results in significant energy being required to effectively jam the system due to the integration gain achieved by the target echos. To successfully attack the system using reasonable power levels, the deterministic components of the signal need to be attacked in such a way that the inherent integration gain of the system can be leveraged.

7.2.2 Mismatched Filtering - Full Pilot Attack

The most obvious pilot attack is a full pilot attack. As summarised in Table 6.1, the number of active carriers per symbol is 27 841 while the number of pilot carriers is 1 196. The pilot carriers make up roughly 4.3% of each symbol. The integration gain on the pilot carriers is therefore reduced by:

$$
\begin{aligned}
G_{\text{reduction}} &= 10 \cdot \log_{10}\left(\frac{\text{All Pilots}}{\text{Active carriers}}\right) \\
&= 10 \cdot \log_{10}\left(\frac{1196}{27841}\right) \\
&= -13.67 \text{ dB}
\end{aligned}
\tag{7.3}
$$

A full pilot attack is demonstrated by injecting it into the surveillance channel with a delay of 30 km and a Doppler shift of 35 Hz. The JSR referenced to the surveillance channel is set to 0 dB and the results are demonstrated in Figure 7.4 where peaks can be seen at the desired location as well as at the expected intervals.

Table 7.5: Simulated jammer parameters for Figure 7.4.

Attack	Range Shift [km]	Doppler Shift [Hz]	JSR [dB]
Full Pilot	30	35	0

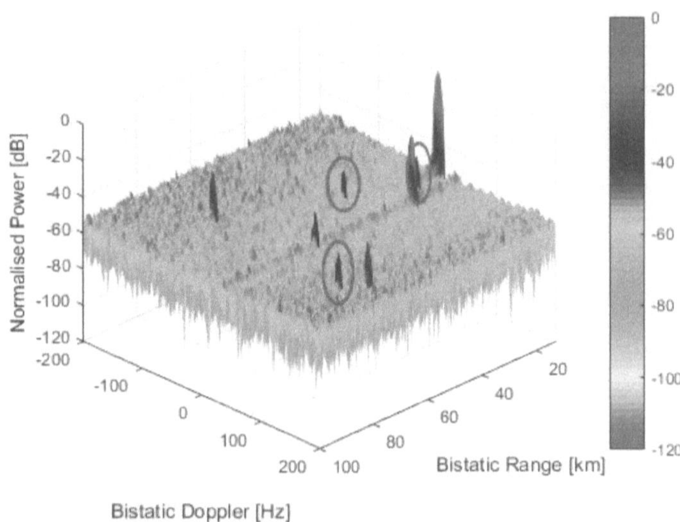

Figure 7.4: Mismatched filtering - Single full pilot attack with pilot map inserted with a Doppler of 35 Hz and a range shift of 30 km.

As expected, due to the reduced integration gain of 13.67 dB and the normalisation reduction of 4.86 dB, the main peak from the pilot pattern appears at -18.46 dB.

Comparing the full pilot attack to noise jamming, it is clear that pilot jamming has the greatest impact on the system. The pilot levels can be adjusted to be near the target level to mask targets at known locations within the ARD map. This is demonstrated in Figure 7.5 where the pilot pattern is placed near a simulated target at 140 Hz in Doppler and 67.4 km in range with a JSR of -10 dB. The ARD map is calculated to 300 km and 300 Hz in range and Doppler respectively to demonstrate the effectiveness of the attack signal in masking a target. The false targets were clearly indistinguishable from the real targets.

Table 7.6: Simulated jammer parameters for Figure 7.5.

Attack	Range Shift [km]	Doppler Shift [Hz]	JSR [dB]
Full Pilot	67.4	140	-10

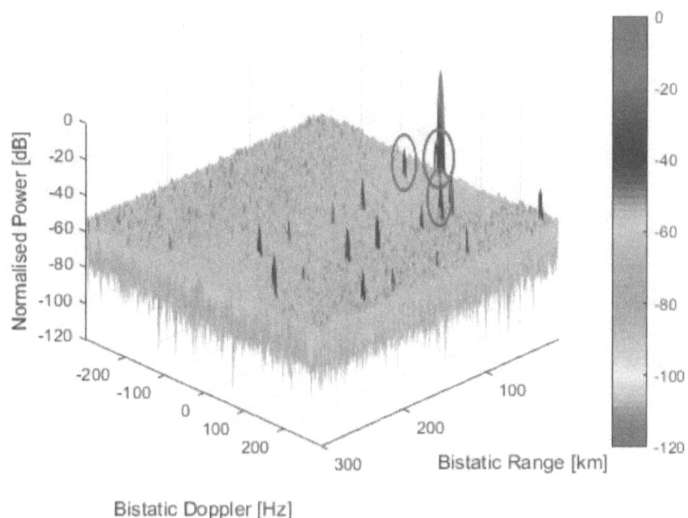

Figure 7.5: Mismatched filtering - Single full pilot attack -10 dB JSR where the attack is placed near a target.

Since an attacker would not necessarily know the exact location of the PR, placing false targets at particular locations within the ARD map via a full pilot attack is not always possible. If a self protection jammer was to be used, the target will be able to receive the DVB-T2 signal and attack a potential PR for a given range since this range will be relative to where the target is. To account for different potential Doppler shifts, the jammer could transmit the attack signal with different Doppler shifts ranging from the minimum to maximum potential Doppler variations of the target. This produces peaks which jam particular range bins while a ridge is formed from each adjacent Doppler shift as demonstrated in Figure 7.6.

Table 7.7: Simulated jammer parameters for Figure 7.6.

Attack	Range Shift [km]	Doppler Shift [Hz]	JSR [dB]
Full Pilot	30	-200 to 200 in 10 Hz intervals	0

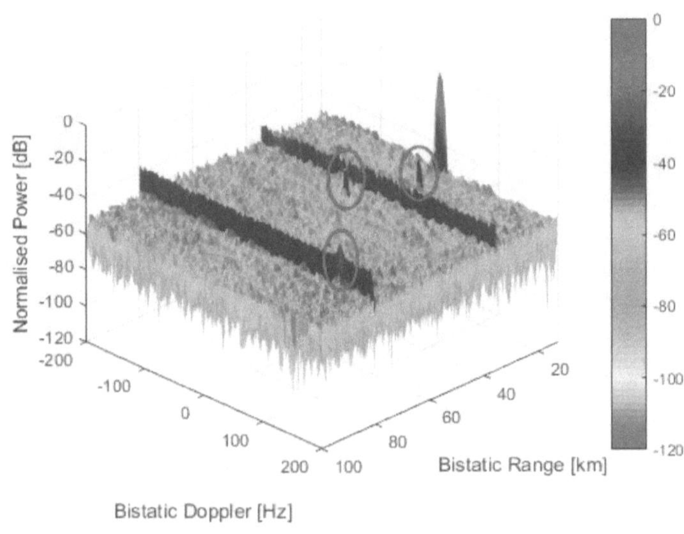

Figure 7.6: Mismatched filtering - Doppler shifted ridges.

Extending the ARD map to illustrate the effectiveness of these ridges across a greater range and Doppler extent is demonstrated in Figure 7.7.

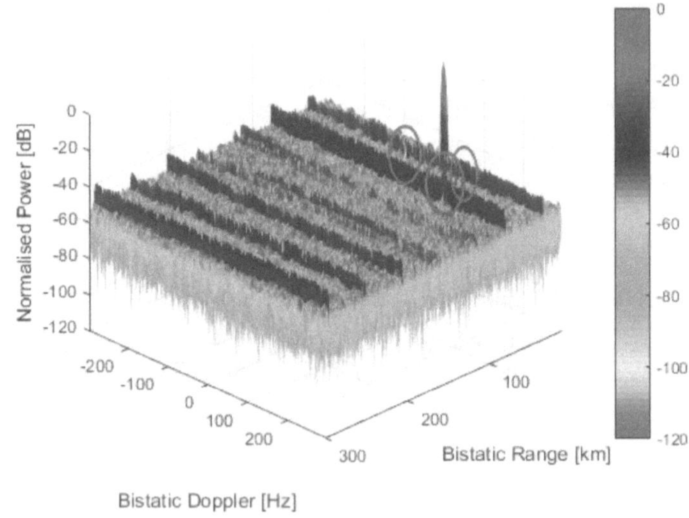

Figure 7.7: Mismatched filtering - Doppler shifted ridges across wider range.

It is clear from Figures 7.6 and 7.7 where the targets are highlighted in red ovals, that ridges can be placed along all Doppler at fixed ranges. The downside to this approach

Table 7.8: Simulated jammer parameters for Figure 7.7.

Attack	Range Shift [km]	Doppler Shift [Hz]	JSR [dB]
Full Pilot	30	-300 to 300 in 10 Hz intervals	0

is that the energy gets spread across the Doppler profile and a higher JSR is needed to obtain the same ridge level for wider Doppler extents. The spacing of the peaks within the ridges can be increased by using larger Doppler shift intervals however, this is a trade-off between completely masking the target and jammer power levels. For example, having peaks spaced every 5 Doppler bins rather than every Doppler bin would require less jammer energy while still resulting in the target being undetected.

7.2.3 Mismatched Filtering - Continual Pilot Attack

The simplest method to attack the PR is to use only the CP carriers since they remain constant from symbol to symbol. The attacker therefore only needs to synchronise to the symbol clock. Attacking the PR using only the CP can cause similar effects to those demonstrated in Figure 6.3 where constant Doppler ridges were formed as the CP remain on the same carriers across symbols. Where the CP and the SP appear on the same carrier, the CP takes the value of the SP. This causes periodicity in time and therefore range which can be seen in Figure 6.3 where sharp peaks appear on top of the ridges.

Unlike when using all the pilots, using only the CP as an attack signal results in an effective bandwidth and therefore integration gain reduction of:

$$
\begin{aligned}
G_{\text{reduction}} &= 10 \cdot \log_{10}\left(\frac{\text{Continual pilots}}{\text{Active carriers}}\right) \\
&= 10 \cdot \log_{10}\left(\frac{178}{27841}\right) \\
&= -21.9 \text{ dB}
\end{aligned}
\tag{7.4}
$$

The large reduction in integration gain compared to the full pilot attack combined with the normalisation process which causes a further reduction of 8.52 dB as shown in (7.1), leads to lower effectiveness of the CP attack. The ridges caused by the CP remaining on the same carrier throughout the frame are seen in Figure 7.8.

The CP can be seen across all range bins as expected but with greatly reduced levels with the peak sitting -29.13 dB below the direct signal peak. While this is the easiest attack to implement, it is also the least effective due to the significantly lower integration gain achieved by the CP.

Table 7.9: Simulated jammer parameters for Figure 7.8.

Attack	Range Shift [km]	Doppler Shift [Hz]	JSR [dB]
CP Only	30	35	0

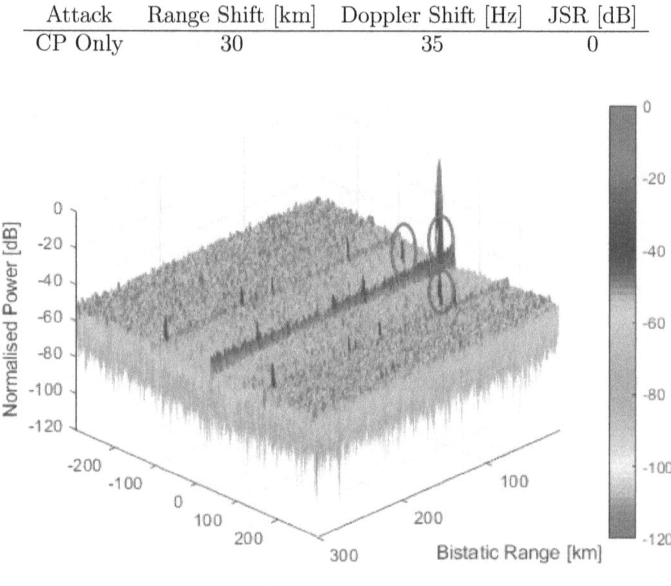

Figure 7.8: Mismatched filtering - Jamming using CP only.

7.2.4 Mismatched Filtering - Scattered Pilot Attack

The SP pattern has the greatest effect on the ambiguities in the ARD map as the SP contain most of the pilot energy. Referring to Table C.3 in Appendix C, there were 27 841 active carriers per 32 768 or 32K extended (32K E) data symbol. Of this, 1 196 were pilot carriers (containing CP and SP) with 1 159 of those being SP. This results in an integration gain reduction of:

$$
\begin{aligned}
G_{\text{reduction}} &= 10 \cdot \log_{10} \left(\frac{\text{Scattered pilots}}{\text{Active carriers}} \right) \\
&= 10 \cdot \log_{10} \left(\frac{1159}{27841} \right) \\
&= -13.81 \text{ dB}
\end{aligned}
\tag{7.5}
$$

As a result of the relatively low bandwidth reduction, and the periodicity of the SP pattern in both fast and slow-time, SP jamming has the most impact and is the most difficult to remove from the ARD map, even with the normalisation process further decreasing their effect by an additional 4.86 dB as shown in (7.2). Figure 7.9 illustrates the effect of using the SP only as an attack signal where the SP can be seen -18.67 dB below the peak signal.

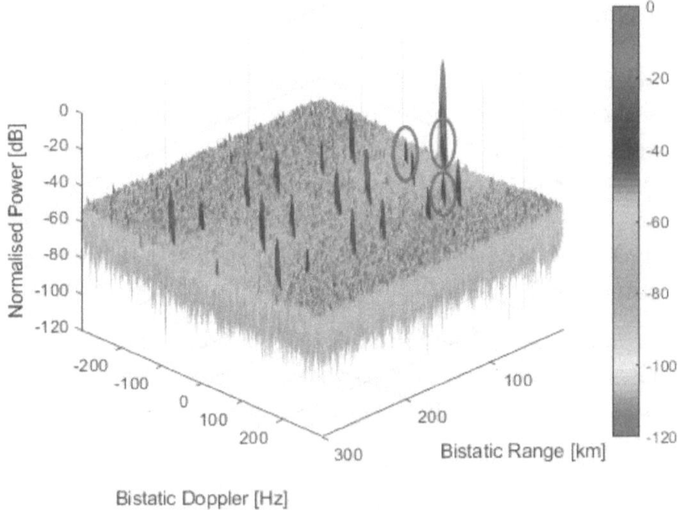

Figure 7.9: Mismatched Filtering - Jamming using SP only.

Table 7.10: Simulated jammer parameters for Figure 7.9.

Attack	Range Shift [km]	Doppler Shift [Hz]	JSR [dB]
SP Only	30	35	0

It is clear that the most effective form of pilot jamming a system using mismatched filtering is a full pilot attack. This is due to the processing gain achieved by the jamming signal. A CP only attack is the easiest to implement since the jammer only has to synchronise with the reference signal and transmit tones on the same carrier without hopping from carrier to carrier between symbols. The CP only attack, however, has the smallest effect on the output ARD map due to its low integration gain.

Since the levels of the attack signal scales directly with the JSR used, the ambiguous peaks caused will scale accordingly. Provided these peaks were above the target level, they will be effective as a means of spoofing or masking a real target. This means that in order to mask a target using a full pilot attack, the JSR needs to be approximately 18.5 dB above the target level within the channel. Since targets can appear as low as 80-90 dB below the direct signal [4, 54], the JSR can be as low as -70 dB and any potential targets will still be masked by additional false targets due to the attack signal.

7.2.5   Mismatched Filtering - Pulse Jamming

A periodic structure in the Doppler domain can be achieved by modulating the attack signal periodically with the OFDM symbol clock. Either the signal can be turned off for every k-th OFDM symbol or only be turned on for every k-th symbol. In both cases this pulse modulation will lead to a periodic structure in the Doppler domain. Since the continuous attack signal using all pilots already produces a set of equi-distant peaks in the Doppler domain, the regular pulse-modulation will combine with these peaks as illustrated in Figure 7.10.

The resultant integration gain achieved is lowered proportional to the number of symbols that the signal remains off. To demonstrate this effect, the result of toggling the full pilot attack signal on every 10-th symbol is shown in Figure 7.10.

Table 7.11: Simulated jammer parameters for Figure 7.10.

Attack	Range Shift [km]	Doppler Shift [Hz]	JSR [dB]
All Pilots Pulsed Every 10-th Symbol	30	35	0

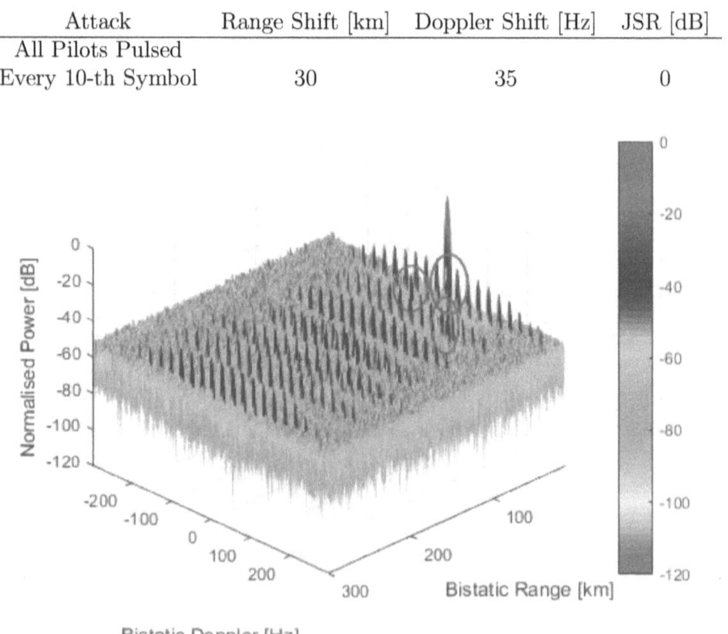

Figure 7.10: Mismatched Filtering - Full pilot toggled on every 10-th symbol.

The integration gain in this case is:

$$G_p = 10 \cdot \log_{10}(N_{active} \cdot C_s \cdot N_{symbols} \cdot T_u)$$
$$= 10 \cdot \log_{10}(1\,196 \cdot 279\text{Hz} \cdot 60/10 \cdot 3.584 \text{ ms}) \tag{7.6}$$
$$= 38.6 \text{ dB}$$

Combining this with a further 4.9 dB reduction due to the normalisation process, the attack signal experiences 33.7 dB of integration gain which puts the peak of the attack signal at - 28.6 dB (33.7 dB - 62.3 dB) which is close to the - 29.1 dB found in Figure 7.10.

It was demonstrated in [135] that toggling the attack signal on for a single symbol results in ridges appearing across Doppler when using inverse filtering. This is due to the manner in which inverse filtering produces an ARD map where a FFT is applied across each symbol to obtain the Doppler spectrum. Applying a large impulse to a single symbol therefore results in a large disturbance in Doppler. With mismatched filtering however, since the slow-time FFT is performed on a data matrix containing matched filtered responses, the energy contained in a single symbol is spread out over the entire range profile. This results in a slight increase in the background noise without inducing sharp ridges across Doppler as demonstrated in Figure 7.11.

Table 7.12: Simulated jammer parameters for Figure 7.11.

Attack	Range Shift [km]	Doppler Shift [Hz]	JSR [dB]
All Pilots Toggled ON for Single Symbol	30	35	0

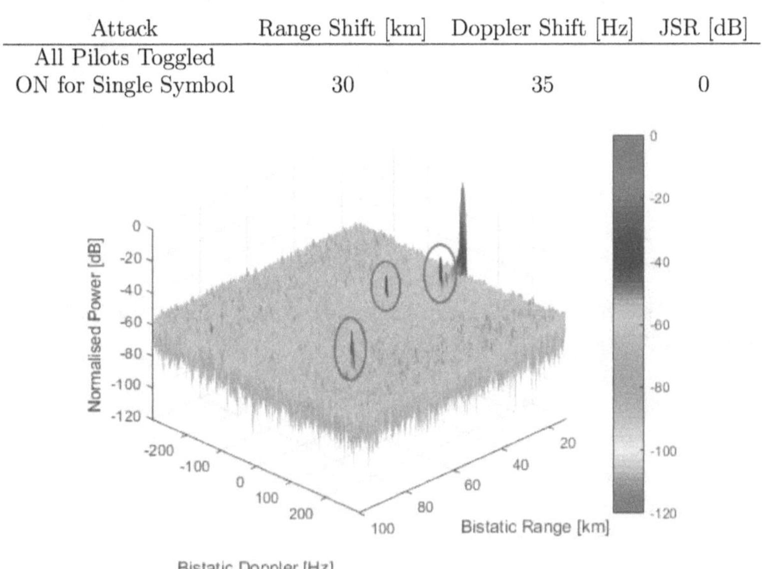

Figure 7.11: Mismatched Filtering - Single symbol attack.

7.2.6   Mismatched Filtering ECCM - Pilot Blanking

To remove any potential pilot jamming, the pilot energy needs to be suppressed in the ARD map. As shown in Section 6.2, the intra-symbol ambiguities were caused by the difference in power between the boosted pilot carriers and the data carriers. As a result, the pilot jamming signal can be removed by blanking the pilots in the remodulated signal at the expense of creating additional known ambiguities.

Figure 7.12 illustrates full pilot jamming processed using a reference with the pilot carriers blanked. As with Figure 7.1, the three simulated targets have been added to the surveillance channel as marked by the red circles.

Table 7.13: Simulated jammer parameters for Figure 7.12.

Attack	Range Shift [km]	Doppler Shift [Hz]	JSR [dB]
All Pilots	30	35	0

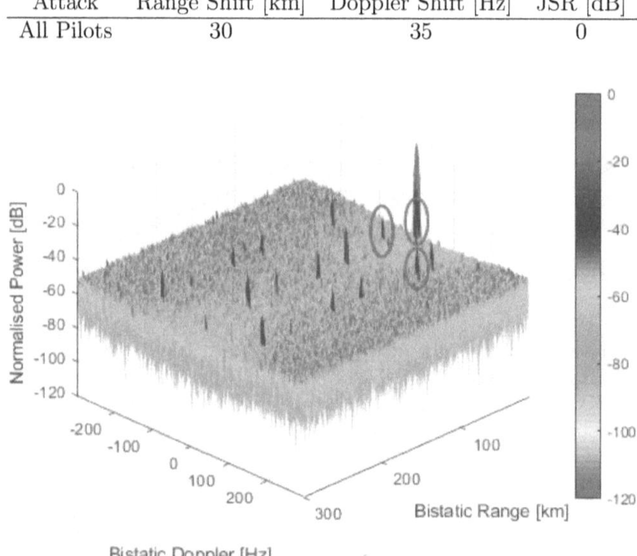

Figure 7.12: ECCM Mismatched Filtering - 3 targets with all pilot jamming using a reference with blanked pilots.

Blanking the pilots in the remodulation stage instead of normalising them results in the removal of the jamming signal as demonstrated in Figure 7.12. All three targets remain, as illustrated by the red ovals, while the jamming signal is removed. As a result of the pilot blanking, new ambiguities were added to the ARD map, however, they appear at known, fixed positions and can therefore be removed by normalising the ARD map based on the expected levels as suggested in [107]. While the noise floor is slightly raised as a result of the jamming signal (compared to Figure 7.13), no target induced ambiguities were present above the noise floor. These ambiguities are clearly seen in Figure 7.13 where a clean surveillance with no jamming and 3 targets is processed using a reference with blanked pilots.

Table 7.14: Simulated jammer parameters for Figure 7.12.

Attack	Range Shift [km]	Doppler Shift [Hz]	JSR [dB]
None	-	-	-

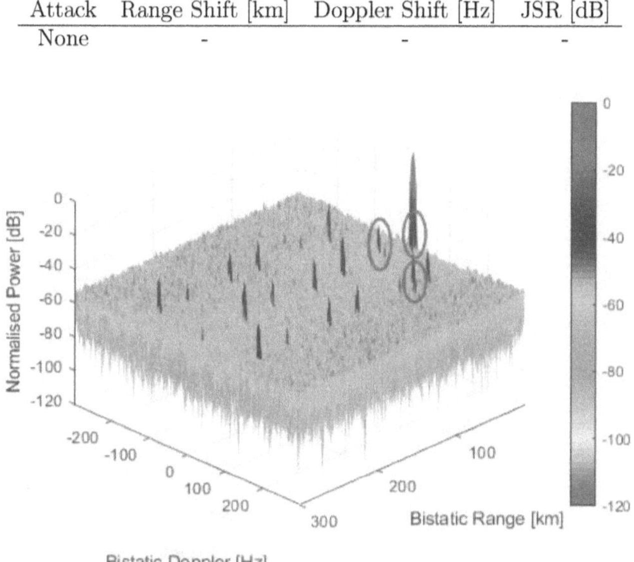

Figure 7.13: ECCM Mismatched Filtering - 3 targets without jamming using a reference with blanked pilots.

7.3 Electronic Attacks using Inverse Filtering

To demonstrate the effects of the different jamming techniques when inverse filtering is used, a reference inverse filter ARD map containing the three simulated targets is demonstrated in Figure 7.14. The target at 140 Hz Doppler shift is seen to wrap around to -118.5 Hz as a result of the Doppler ambiguity issues outlined in Section 6.4.2. The target at 25 km in range and 20 Hz Doppler is seen to spread out in the Doppler dimension due to it straddling between two adjacent Doppler bins, resulting in a peak reduction of 3 dB. The noise floor is seen to be at approximately -85 dB.

It is important to note that the noise floor as it appears in Figure 7.14 is an artificial noise floor rather than a system noise floor. This is partly due to the simulated nature of the surveillance channel but more importantly due to the symbol mismatches caused by the inserted targets. As inverse filtering is performed using element-wise division, if the surveillance channel contains the exact content that is found in the reference channel, that is to say no targets, Doppler shifts or system noise, the noise floor at non-zero Doppler values becomes undefined. As such, any non zero-Doppler within the surveillance channel results in inter-carrier interference that leads to additional noise being present in the output ARD map.

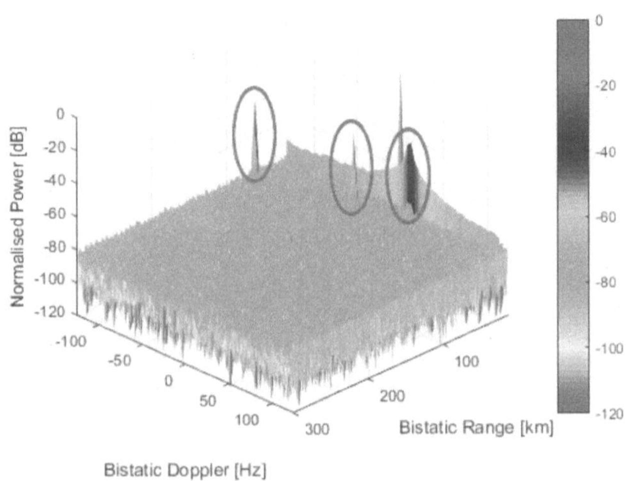

Figure 7.14: Inverse filtering benchmark performance with 3 targets and no jamming.

Table 7.15: Simulated jammer parameters for Figure 7.14.

Attack	Range Shift [km]	Doppler Shift [Hz]	JSR [dB]
None	-	-	-

The spreading of targets in the Doppler dimension is a result of the target response straddling between Doppler bins and can be largely removed through windowing as demonstrated in Figure 7.15 where a Blackman window is applied.

Table 7.16: Simulated jammer parameters for Figure 7.15.

Attack	Range Shift [km]	Doppler Shift [Hz]	JSR [dB]
None	-	-	-

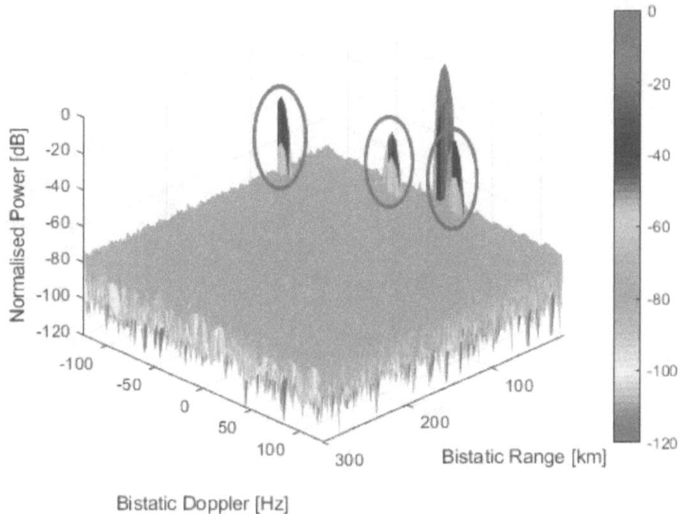

Figure 7.15: Inverse filtering benchmark performance with Blackman window.

As expected, the target sidelobes were reduced at the expense of widening the main lobe, especially for the targets that dont straddle two bins. For the remainder of the results, no windowing is used unless otherwise stated. Ambiguous target unwrapping will not be discussed in this work as it is not within the scope.

7.3.1 Inverse Filtering - Noise Jamming

As mentioned in section 7.3, the noise floor of the ARD map is highly dependant on the amount of noise on each carrier within the surveillance channel. As a result, when no noise is present (only the direct signal and target echos), the noise floor is well below the noise floor of the system when using mismatched filtering. This artificial noise floor changes drastically with the addition of noise to the system. Figure 7.16 demonstrates the effect of adding Gaussian white noise to the system with a JSR of 0 dB.

Table 7.17: Simulated jammer parameters for Figure 7.16.

Attack	Range Shift [km]	Doppler Shift [Hz]	JSR [dB]
AGWN	-	-	0

When comparing the results demonstrated in Figure 7.16 to the results demonstrated when mismatched filtering is used, it is clear that the noise floor of the inverse filtering

116

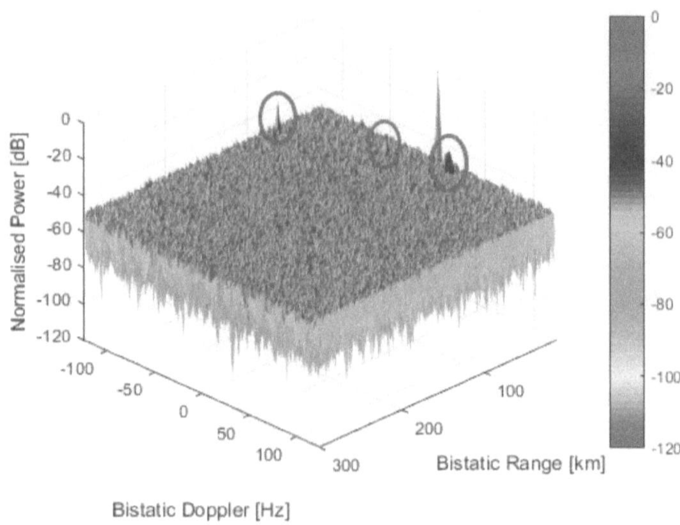

Figure 7.16: Inverse Filtering - Noise jamming 0 dB JSR.

ARD map is significantly effected. The noise floor for both the mismatched filtering and inverse filtering ARD maps in Figures 7.2 and 7.16 respectively were G_p below the zero-range, zero-Doppler peak.

Increasing the noise JSR to 10 dB results in the targets being masked completely as demonstrated in Figure 7.17.

Table 7.18: Simulated jammer parameters for Figure 7.17.

Attack	Range Shift [km]	Doppler Shift [Hz]	JSR [dB]
AGWN	-	-	10

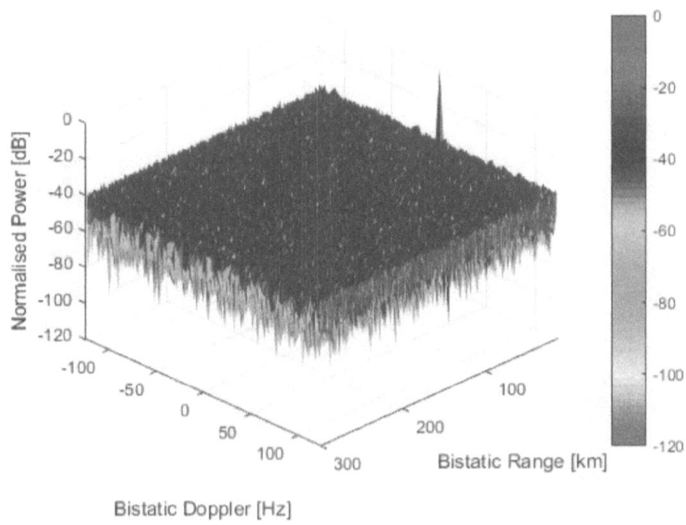

Figure 7.17: Inverse Filtering - Noise jamming 10 dB JSR.

It is clear that inverse filtering is not as robust against noise jamming as mismatched filtering as the difference in noise floor levels at 3 dB JSR is almost 5 dB between mismatched filtering and inverse filtering. This trend continues as the JSR increases to 10 dB as can be seen in Figures 7.3 and 7.17. Unlike the matched filtering approach which is an optimal filter for extraction of signals, the inverse filter is highly sensitive to disturbances in the surveillance channel. This phenomenon results from the element wise division step in the inverse filtering process. When the surveillance channel is normalised by the reference channel, any additional noise that is present in the surveillance channel that is not present in the reference channel does not get normalised. These errors in each carrier were therefore present as randomly rotating phasors which when the DFT is applied, results in random noise (i.e. an elevated noise floor) across the Doppler dimension as seen in Figure 7.17.

7.3.2 Inverse Filtering - Full Pilot Attack

A full pilot attack on inverse filtering is presented where the attack signal has a delay of 30 km and a Doppler shift of 35 Hz. The JSR referenced to the surveillance channel is 0 dB and the results are demonstrated in Figure 7.18.

Table 7.19: Simulated jammer parameters for Figure 7.18.

Attack	Range Shift [km]	Doppler Shift [Hz]	JSR [dB]
All Pilots	30	35	0

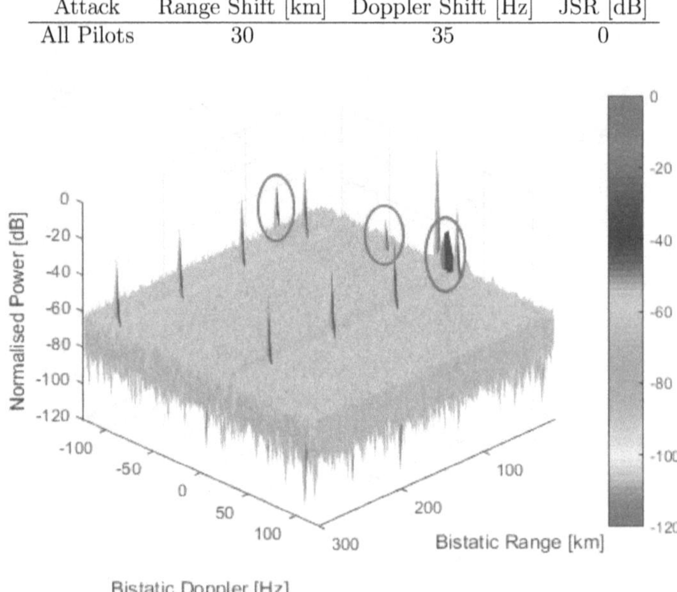

Figure 7.18: Inverse filtering - Full pilot attack with 0 dB JSR.

The resultant peaks represent a shifted version of the pilot pattern where the first peak appears at a range of 30 km and a Doppler shift of 35 Hz along with a constant Doppler ridge. These peaks were large enough to be detected as false targets within the ARD map. A ridge across range at constant Doppler can be seen at a lower level to the peak. This ridge is caused as a result of the continual pilots remaining on a constant carrier from symbol to symbol.

As with mismatched filtering, the effect of the attack signal is reduced during the division step of inverse filtering (where the pilot carriers were effectively normalised) which results in a further 4.9 dB loss in integration gain for the SP pattern. The end result is peaks sitting at -18.9 dB $(13.7 + 4.9)$. Using a full pilot attack, one can place these peaks at any desired position, with each additional peak resulting in only a 3 dB drop in peak power level for the same JSR. This is demonstrated in Figure 7.19 where two pilot patterns were inserted 35 Hz apart from each other (at 35 Hz and 70 Hz respectively). As was the case with mismatched filtering, this approach is significantly more effective than barrage noise jamming due to the added integration gain achieved by the pilot signal.

Table 7.20: Simulated jammer parameters for Figure 7.19.

Attack	Range Shift [km]	Doppler Shift [Hz]	JSR [dB]
All Pilots	30	35 & 70	0

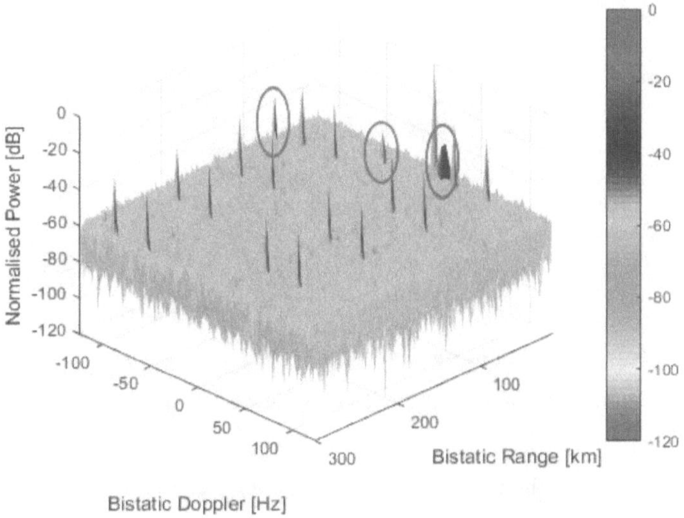

Figure 7.19: Inverse Filtering - Full pilot attack with two attack signals 0 dB JSR

7.3.3   Inverse Filtering - Continual Pilot Attack

Using only the continual pilots to attack the PR is the simplest form of attack since the pilots remain on the same carrier from symbol to symbol. Due to the lack of periodicity in slow-time, ridges were produced at constant Doppler across range as demonstrated in Figure 7.20.

Table 7.21: Simulated jammer parameters for Figure 7.20.

Attack	Range Shift [km]	Doppler Shift [Hz]	JSR [dB]
CP Only	30	35	0

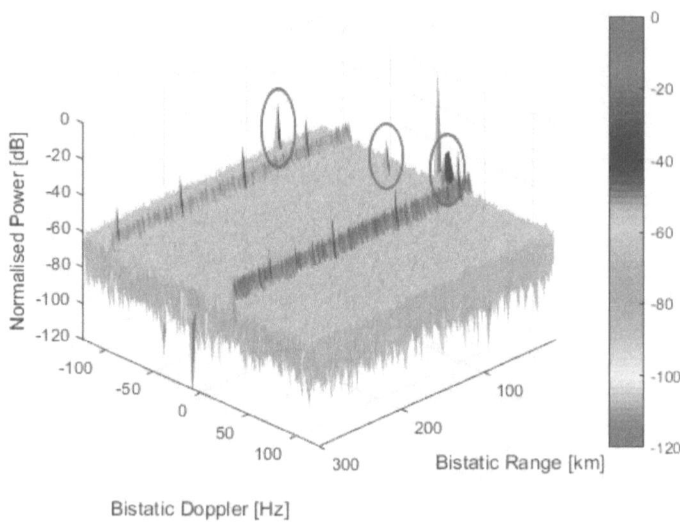

Figure 7.20: Inverse Filtering - Continual pilot attack 0 dB JSR.

The peaks on top of the ridges appear at -41.8 dB while the average power of the ridges is 6 dB lower at -47.8 dB. This is similar to what was seen when using mismatched filtering where a reduction in bandwidth results in a corresponding reduction in integration gain as described in (7.4). The reduction in integration gain along with the reduction in power as a result of normalisation (where the surveillance channel is divided by the reference pilots) during the inverse filtering process results in an overall level reduction of 30.4 dB.

The small peaks on top of the ridges occur where the scattered pilots appear on the same carrier as the continual pilots as the resultant pilot values were set to that of the scattered pilots, resulting in a periodic pilot response on that carrier. If this is removed from the pilot pattern and only a CW tone is transmitted, those peaks will be removed, leaving only the constant Doppler ridges.

7.3.4   Inverse Filtering - Scattered Pilot Attack

For completeness, we consider only using the scattered pilots as an attack signal. The resultant effect is similar to the effect of a full pilot attack as demonstrated in Figure 7.21. The scattered pilots alone create peaks without the constant Doppler ridges caused by the continual pilots as demonstrated in Figure 7.20.

Table 7.22: Simulated jammer parameters for Figure 7.21.

Attack	Range Shift [km]	Doppler Shift [Hz]	JSR [dB]
SP Only	30	35	0

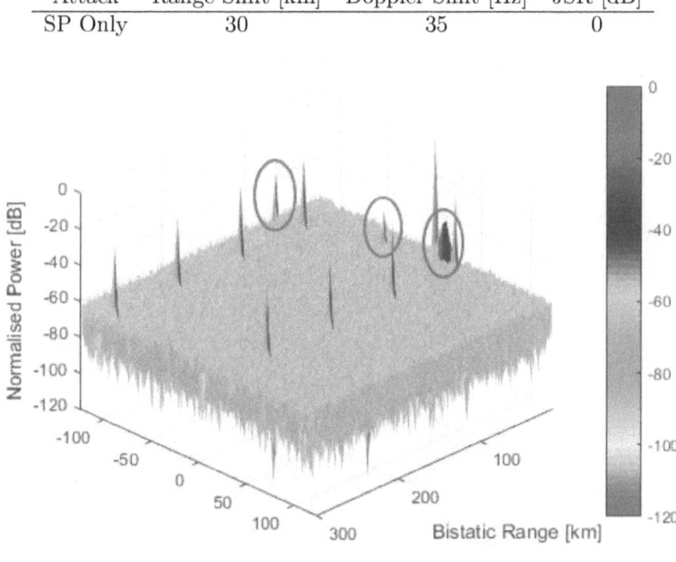

Figure 7.21: Inverse Filtering - Scattered pilot attack 0 dB JSR.

This type of attack is not favourable as synchronisation of the complete pilot map is required to carry out the attack, in which case a full pilot attack would be preferred.

7.3.5 Inverse Filtering - Pulse Jamming

Continuing from the pulsed attack that was performed on the system utilising mismatched filtering, the effects of pulse jamming using inverse filtering were investigated. The additional Doppler-domain peaks that are induced through pulsed jamming can be explained in the following way.

In the Fourier domain, the modulation can be understood as a multiplication of two pulse trains with different numbers of peaks on the same grid. In the dual domain this is equal to the convolution of two pulse trains with the same number of peaks as their Fourier counterparts. Let p_1 and p_2 be the number of peaks in the first and second signal, respectively. To find the number of peaks in the resulting signal, we have to first find the largest common denominator, d, for each peak. The number of peaks in the convolved signal is then given by $1/(d \cdot p_1 \cdot p_2)$.

In this case, one signal always has 2 peaks in the Doppler domain meaning that $d \in \{1, 2\}$. If we turn the attack signal on or off every k-th symbol, we get $2 \cdot k$ peaks if k is odd. If k is even we get k peaks. Pulse jamming where the attack signal is toggled on every 10-th symbol is demonstrated in Figure 7.22. The JSR of the attack is 0 dB whereby the average power in the attack signal is the same as the average power in the surveillance channel.

Table 7.23: Simulated jammer parameters for Figure 7.22.

Attack	Range Shift [km]	Doppler Shift [Hz]	JSR [dB]
All Pilots Pulsed			
Every 10-th Symbol	30	35	0

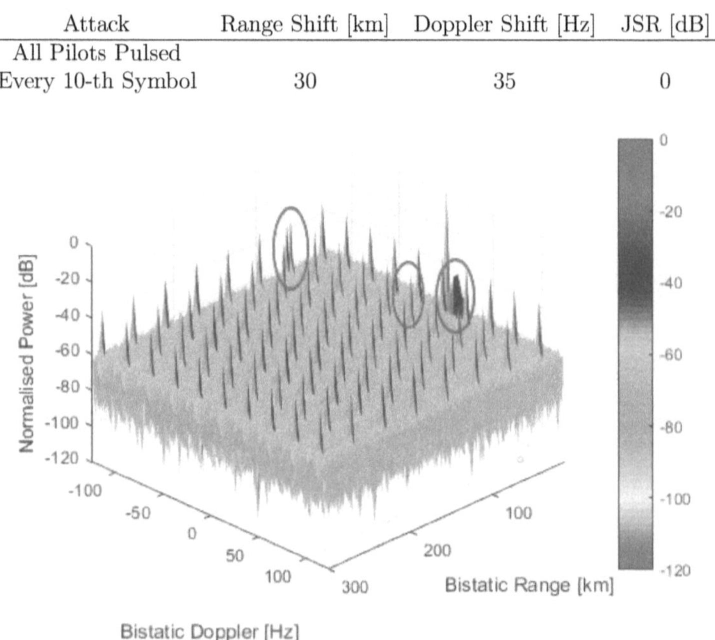

Figure 7.22: Inverse Filtering - Pulsed jamming toggled on every 10-th symbol

Figure 7.22 illustrates an ARD map where a pulsed attack signal consisting of the full pilot signal with relative range of 30 km and Doppler of 35 Hz has been applied. The attack signal is turned on every 10-th symbol which leads to recurring peaks in the Doppler domain. As demonstrated, the periodicity of the CP contribution, which only appears once per delay bin, is multiplied by the factor k while the SP contributions also get multiplied by k, which in this case is 10.

In the extreme case of $k = M$, one single symbol per CPI is attacked. This leads to solid ridges over the Doppler domain for constant delay which is demonstrated in Figure 7.23. The spacing of these ridges is given by the spacing of the range ambiguity function of the pilot signal. Due to the reduced duty cycle, we lose $20 \cdot \log_{10}(M)$ of peak level in the ARD map which amounts to 35.5 dB for $M = 60$. Note that a JSR of 0 dB in this case implies that the attack signal is at same average power as the surveillance signal for only this one symbol. This means that the average power needed by the jammer is lowered by $10 \cdot \log_{10}(M) = 17.8$ dB compared to continuous jamming.

Table 7.24: Simulated jammer parameters for Figure 7.23.

Attack	Range Shift [km]	Doppler Shift [Hz]	JSR [dB]
All Pilots Toggled ON for Single Symbol	30	35	0

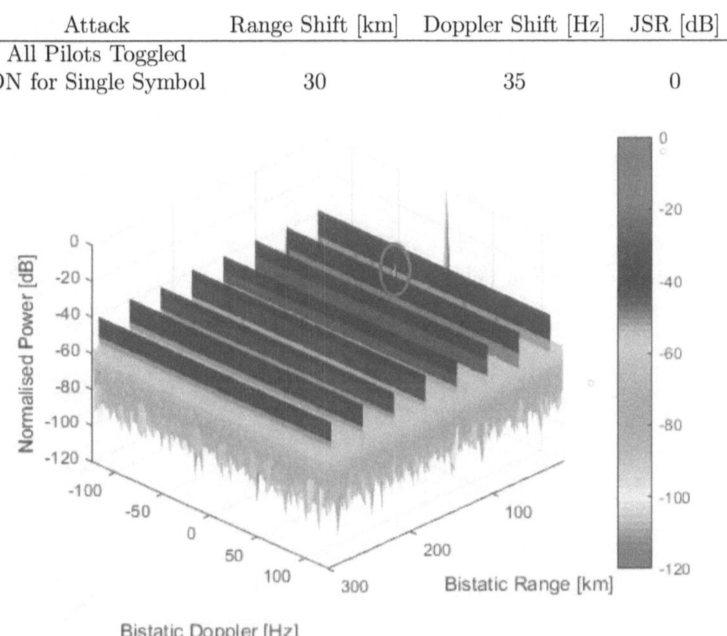

Figure 7.23: Inverse Filtering - Single symbol jamming 0 dB JSR.

Due to the nature of the radar system, the receiver location is assumed to be unknown. As a result, the exact distance to the receiver and velocity of the target relative to the receiver is also unknown. The target will, however, be able to synchronize itself with the transmit signal and transmit the jamming signal at the same time as the incident signal reaches the target. Both signals will therefore arrive at a potential receiver concurrently as both the target echo and jamming signal will have the same distance to travel to the receive site.

This is very attractive for self-protection jamming even when the location of the receiver is unknown. A target can therefore receive the DVB-T2 signal utilised by the PR and transmit a jamming signal that only attacks the delay bin of the target. For this, even very low power is sufficient due to the processing gain. It must, however, be noted that regardless of the transmitted power level, a sophisticated ES receiver could potentially locate the self-protection jammer.

7.3.6   Inverse Filtering ECCM - Pilot Boosting

To mitigate the effect of the attack signals, the pilot energy needs to be suppressed in the ARD map. Due to the way inverse filtering is performed, the pilot signals can be suppressed through the normalisation process which, unlike with mismatched filtering which normalises through convolution, inverse filtering normalises through division. In the same way that the pilots were normalised by dividing the attack signal by the boosted pilot levels in the reference signal, the higher powered attack signal can be normalised further by boosting the pilot signals within the reference signal.

Adding a 20 dB boost to the pilot levels results in the pilot attack signal being suppressed as demonstrated in Figure 7.24.

Table 7.25: Simulated jammer parameters for Figure 7.24.

Attack	Range Shift [km]	Doppler Shift [Hz]	JSR [dB]
All Pilots	30	35	0

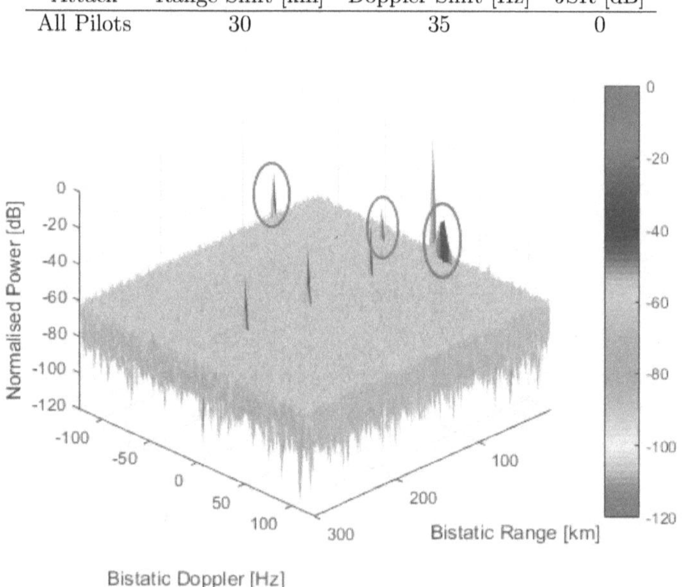

Figure 7.24: ECCM Inverse Filtering - Full pilot jamming with 20 dB reference pilot boost.

Similar to when pilots were blanked in mismatched filtering, when using boosted pilots in inverse filtering, additional ambiguities were inserted into the ARD map. The ambiguous peaks were however, much less of a problem in inverse filtering when compared to mismatched filtering as these peaks simply appear along the zero-Doppler profile rather than at different positions in Doppler.

The boosting of pilots also has the unwanted effect that if a target appears within the ARD map, the target will have periodic ambiguities which mirror those seen along the zero-Doppler line, albeit at significantly lower levels as demonstrated in Figure 7.25.

Table 7.26: Simulated jammer parameters for Figure 7.25.

Attack	Range Shift [km]	Doppler Shift [Hz]	JSR [dB]
None	-	-	-

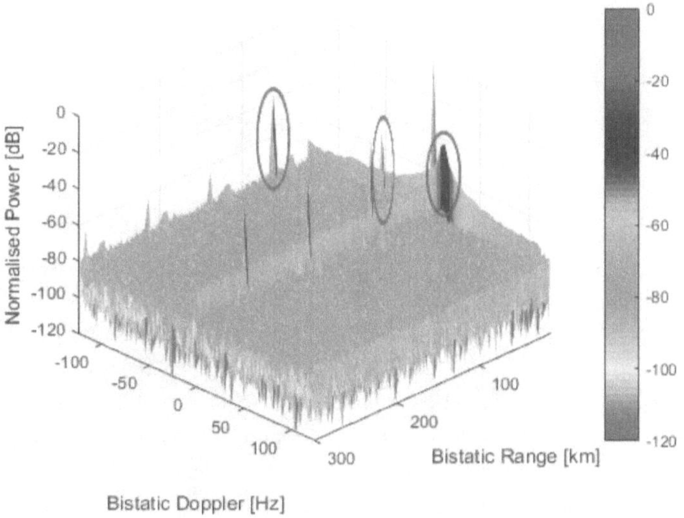

Figure 7.25: ECCM Inverse Filtering - No jamming with 20 dB reference pilot boost.

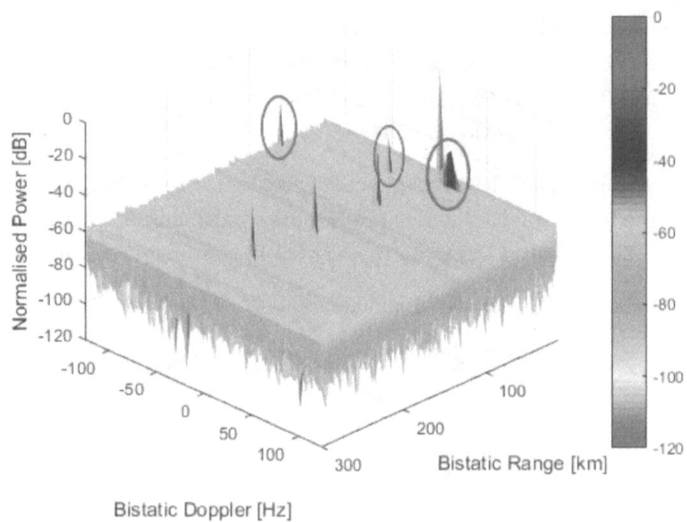

Figure 7.26: ECCM Inverse Filtering - Single symbol jamming with 20 dB reference pilot boost.

Table 7.27: Simulated jammer parameters for Figure 7.26.

Attack	Range Shift [km]	Doppler Shift [Hz]	JSR [dB]
All Pilots Toggled ON for Single Symbol	30	35	0

The boosted pilot technique also minimises the effect of pulsed and single symbol jamming to a large extent as demonstrated in Figure 7.26 where the sharp ridges were shown to be removed. However, the increase in noise floor is still significant and the noise profile is no longer Gaussian in nature. Ideally, the use of boosted pilots will only be to counter jamming and should therefore be dynamically adjusted based on the system requirements.

7.4 Discussion of Results

A brief discussion comparing the results of jamming a system using mismatched filtering and inverse filtering is presented.

7.4.1 DVB-T2 Noise Jamming

Broadband noise jamming has been applied to both processing techniques where it was demonstrated that mismatched filtering is more robust to noise than inverse filtering.

Since noise jamming is a linear process whereby the energy within the signal is spread across the entire ARD map without any integration gain, it is clear that to have any meaningful effect on the system, the noise power required is significant. An ARD map processed using mismatched filtering sees a meaningful increasing in background noise level when the JSR approaches 0 dB while the same ARD map using inverse filtering experiences a significant increase in background noise level at around -5 dB JSR.

These levels were still significant, however since the jammer would require power levels roughly equal to or -5 dB below the reference transmitter to mask target echos -35 dB below the zero-range, zero-Doppler peak.

7.4.2 DVB-T2 Full Pilot Attack

Using the pilot carriers as an attack signal is very favourable for two reasons, firstly, unlike noise jamming where no integration gain is achieved, significant integration (in the order of 48 dB) is achieved. Secondly, using the pilot carriers as an attack signal also benefits from the fact that it is more difficult to detect a signal that is synchronised with the reference signal than one that is not due to it being masked by the desired, high powered reference signal.

Using a full pilot attack results in significant ambiguities appearing in the ARD map regardless of the processing technique used. A full pilot attack has been demonstrated to be an effective means of attacking a DVB-T2 PR.

7.4.3 DVB-T2 Continual Pilot Attack

Using only the continual pilots, the attacker gives up the periodicity in slow time as the pilots remain on the same carrier for the entire CPI. The advantage of this is that the attack signal is simple to execute and leads to ridges across the range profile of the ARD at constant Doppler bins defined by the Doppler shift added by the attacker.

Additionally, the use of continual pilots as an attack signal leads to reduced attack signal integration gain due to the reduced bandwidth within the signal. For a 32K DVB-T2 signal, the integration gain reduction from only using the continual pilots is 21.9 dB. Coupling this with a further reduction in gain of 8.52 dB due to the normalisation process, the continual pilots experience 11.9 dB ($G_{\text{reduction (CP)}} - G_{\text{reduction (Full pilot)}}$ $= (21.9 + 8.52) - (13.67 + 4.86)$) less integration gain than the full pilot attack signal.

7.4.4   DVB-T2 Scattered Pilot Attack

Using the scattered pilots as the attack signal creates an effect similar to that seen when using a full pilot attack with the difference being the lack of ridges across range and decreased integration gain resulting from the loss of continual pilots. For this reason and the fact that using a scattered pilot attack has the same complexity as using a full pilot attack, scattered pilot attacks are not recommended.

7.4.5   DVB-T2 Pulsed Jamming

Since an attacker will in all likelihood have no knowledge of the PR receiver location, creating ridges across Doppler at specific ranges is a desirable attribute for an attack signal. To create these ridges across Doppler, slow-time periodicity needs to be added to the signal. To achieve this, the full pilot attack signal is toggled on every k-th symbol.

It has been demonstrated that toggling the attack signal on every 10-th symbol results in ambiguities that spread across Doppler. Toggling the signal on for a fraction of the CPI also results in reduced continual power requirements due to the reduced duty cycle.

It is also demonstrated that transmitting the attack signal for a single symbol causes severe ridges across Doppler when using inverse filtering. This is due to the fact that the system is essentially experiencing an impulse in one of the symbols before a Doppler FFT is performed. This effect is not replicated when mismatched filtering is used due to the energy being spread across the ARD map during the convolution process. A minimum of 4 symbols out of the 60 (15%) are need to be attacked to cause noticeable effects in the mismatched filtering process.

7.4.6 DVB-T2 Counter-counter Measures

A brief discussion of counter-counter measures has been provided. To remove the attack signal from the system when using mismatched filtering, the pilot signals can be blanked. While this removes the ambiguities resulting from the pilot attacks, it also adds additional ambiguities to the standard ambiguity function. These ambiguities are however, deterministic and can therefore be removed through additional processing as discussed in [107].

Counter-counter measures for inverse filtering have also been demonstrated where the pilot attacks were normalised using boosted pilots in the remodulated reference signal. Boosting the pilots in the reference signal leads to additional ambiguous peaks within the ambiguity function however, these peaks only appear along the zero-Doppler line and were therefore easily dealt with. Depending on the power level of the target echos, the additional ambiguities created as a result of the boosted pilots could potentially appear above the background noise floor, however, this is unlikely.

It is noted that these ECCM techniques should be dynamic and only be used when jamming is detected.

7.5 Chapter Summary

The performance of both mismatched filtering and inverse filtering processing techniques has been evaluated in the context of a intentional jamming where it has been demonstrated that mismatched filtering is more robust against noise jamming, requiring 5 dB more JSR than inverse filtering, to achieve the same effect. This apparent robustness of mismatched filtering is due to the matched filter convolution step that distributes the noise across the entire range-Doppler profile. In the case of inverse filtering, the more noise there is on each OFDM carrier, the higher the resultant inter-symbol interference and therefore the higher the background noise floor within the ARD map as discussed in Section 7.3.1.

Unlike with FM based PR where the attack signals do not achieve any significant integration gain, the deterministic nature of the DVB-T2 signal allows for more advanced attack signals to be used where significant integration gain can be achieved. As a result, more advanced attack signals have been investigated and the ability to insert false targets with relatively low jamming power has been demonstrated. It is demonstrated that both mismtached filtering and inverse filtering techniques can be successfully attacked using both the continual and scattered pilot patterns. It has been demonstrated that false targets can be placed within the ARD map by synchronising the jammer with the reference signal and then transmitting the pilot pattern with appropriate range and Doppler shifts.

Since the pilot pattern has periodicity in fast time, ridges were created in the range dimension of the ARD map. By pulsing the jamming signal, it is demonstrated that periodicity in fast time can be added, resulting in peaks forming in both the range and Doppler dimensions as demonstrated in Figures 7.11 and 7.22.

A brief discussion on possible counter measures is provided. It is demonstrated that the effects of the attack signals can be almost entirely suppressed. Additional ambiguous peaks appear within the ARD map when the attack pilot signal is suppressed through blanking of the reference pilots in mismatched filtering as demonstrated in Figure 7.12, however, these ambiguities were purely deterministic and can therefore be removed by a technique proposed in [107]. When the attack signal is suppressed through boosted reference pilots (when using inverse filtering), additional ambiguities appear in fast-time on the zero-Doppler line. These zero-Doppler ambiguities were also deterministic and can be removed by blanking the zero-Doppler line. This makes it an attractive method for removing pilot jamming.

Chapter 8

Conclusions and Future Work

The research shows that PR is inherently jammable, often with very low powers when thought is given to the right type of interference required. Nevertheless, a jammer needs to expend adequate effort in terms of jamming-power, as the location of the PR receiver is generally unknown. Furthermore, the jammer, or EA tactic more generally, would not know how effective it is at impairing PR performance. Against conventional radar an additional ES "look-through" asset may be utilised to assess the effectiveness of an EA, but such assessment is difficult to achieve with a PR. Therefore, PR is susceptible to EA, however, it is not an effortless undertaking on the part of the PR countermeasure.

It is also important to note that due to the nature of PR, the number of Tx/Rx pairs and channel variations is massive, making it unfeasible to draw generic conclusions from the results. All objectives that were set at the beginning of this study in Chapter 1 have been completed:

FM Passive Radar

- The effect of the DSI canceller in FM PR in the presence of intentional and unintentional interference has been quantified.

- A complete waveform study has been performed to demonstrate the optimal FM jamming waveform.

- Basic ECCM has been discussed to counter potential jamming of FM PR.

- A representative measurement of a real FM PR has been shown.

DVB-T2 Passive Radar

- Analysis of the two main processing techniques used in DVB-T2 PR has been performed.

- An in depth performance review of the two most common processing techniques in the presence of noise jamming has been presented.

- The deterministic components of the DVB-T2 waveform have been demonstrated to be an effective form of EA for both mismatched filtering and inverse filtering processing techniques.

- Basic ECCM has been presented to counter potential pilot attacks on DVB-T2 PR.

8.1 Conclusions

This book has demonstrated that successful jamming of an FM based passive radar depends on two main factors, the jamming waveform and knowledge of the receiver location. The effect of different jamming waveforms is summarised in Table 4.1 where the most effective jamming waveform is shown to be a tone transmitted on the same centre frequency as the reference signal.

The importance of a DSI canceller has been demonstrated where it has been shown that it not only improves overall performance under normal operation but also acts as a means of suppressing the jamming signal. Due to the DSI canceller, an FM PR is more robust against jamming when the reference and surveillance antennas were co-located where the jamming signal appears in both channels as opposed to being separated where the reference channel remains free of jamming interference. As a result, the DSI canceller can improve the performance of an FM PR by an average of approximately 10% in the presence of jamming when the jamming signal appears in both the reference and surveillance channels of the receiver.

It has been demonstrated that the dimension in which the CFAR filter is applied (either in range or in Doppler), plays an important role in target detection and extraction when tone jamming is applied to FM PR. Under normal operating conditions, the CFAR filter for an FM PR is applied in the Doppler dimension due to the fluctuating range resolution as a result of signal content. FM waveforms have a constant average power, approximated by (3.10). As a result, by decreasing the number of sidebands required to accurately represent the message signal (decreasing β), the carrier tone levels were increased. This increased tone level results in an increase in correlation artefacts in the output ARD maps. As a result of these additional artefacts appearing in the output ARD, the background noise of the system can no longer be modelled as Gaussian and therefore the CFAR filter cannot accurately set a detection threshold, leading to further performance reductions.

Due to the ridges in the ARD maps caused by tone jamming, it is almost impossible for the CFAR filter to detect a target when applied in the Doppler domain. Applying the same CFAR filter in the range dimension allows for higher probability of detection however it comes with the significant drawback that its performance is highly dependant on the average bandwidth over the CPI. It is therefore suggested that in order to effectively detect targets when encountering a possible tone jamming attack, the CFAR filter be applied in both range and Doppler dimensions or switch between the two depending on environmental conditions. A potential counter to tone jamming would be to blank or filter out tones as they appear in the receiver.

In the case of DVB-T2 PR, it has been demonstrated that deterministic components of the OFDM signals can be used by an attacker to jam, spoof or overload the system. The pilot signals can be used to produce false targets in particular range and Doppler bins and due to the integration gain achieved by the pilot jamming, lower power levels can be used than with broadband noise jamming. The two main processing methods employed in DVB-T2 PR have been investigated and their performance in the presence of jamming has been evaluated. It has been shown that mismatched filtering is more robust to noise jamming than inverse filtering due to the robustness of the correlation process.

Along with an in depth analysis of the two main processing techniques for DVB-T2 PR, various attacks have been investigated including noise jamming, full pilot attacks, continual pilot attacks, scattered pilot attacks and pulsed jamming attacks. Broadband noise jamming has been demonstrated to have a linear effect on the system as expected with an increase in system noise floor proportional to the amount of noise injected into the PR frontend. Full pilot attacks were the most complicated to deploy however, they have been demonstrated to be the most effective due to the increased integration gain that is achieved.

Continual pilots have been demonstrated to be the least effective of the pilot attacks however, the continual pilot attack is the simplest attack to implement as the pilots themselves remain on the same carriers from symbol to symbol. This removes the need to jam different carriers on each symbol. Scattered pilot attacks were relatively effective however, since they require the same level of abstract to deploy as full pilot attacks, they are not preferred. The final type of attack, pulsed attacks have been shown to be highly effective against DVB-T2 PR, especially when used with inverse filtering. Toggling a full pilot attack on and off every k-th symbol leads to ridges appearing across Doppler at various range bins which can be altered as desired.

For the purpose of this work, a full pilot attack was toggled every 10-th symbol which demonstrated the effect of these ridges for both mismatched and inverse filtering. Another advantage to utilising a pulsed attack was that the duty cycle was significantly decreased which results in lower continual power requirements for the jammer. Transmitting the attack signal for a single symbol was shown to cause severe ridges across range and Doppler when using inverse filtering however. Applying the same attack to the system when using mismatched filtering demonstrated little effect as the energy was spread across the ARD map as a result of the correlation process. In order to have an effect on a system utilising mismatched filtering, it was deemed that a minimum of 4 symbols out of the 60 (15%) needs to be attacked.

Potential ECCM techniques were briefly investigated where pilot blanking and pilot boosting for mismatched filtering and inverse filtering respectively were demonstrated. Pilot blanking was demonstrated to produce additional ambiguities within the ARD map when processed using mismatched filtering. Using pilot blanking combined with ambiguity removal as suggested in [107], the effects of pilot jamming can be largely mitigated. When using inverse filtering, pilot boosting can be used and as with pilot blanking in mismatched filtering, the ambiguity removal process suggested in [107] can be used to remove any additional ambiguities resulting from the mitigation process.

8.2 Future Work

There is scope for future work in both the FM and DVB-T2 PR configurations with the following presented for consideration.

- Significantly more real world data is required before PR vulnerabilities can be fully statistically quantified for use in both the military and commercial context.

- Given the DVB-T2 signals deterministic nature, experimental results should closely match the simulated results shown in this work however, a real world measurement campaign of DVB-T2 jamming should be carried out to further validate the simulated results.

- The system performance at the output of a tracking algorithm could be evaluated in the presence of jamming. Since the potential number of channel combinations are huge, it is potentially impossible to evaluate every scenario however, this could be done on a system by system basis to evaluate the vulnerabilities of each particular system which could be fed into a vulnerability or threat evaluation database.

- ECCM has been briefly discussed in this work however, a full investigation into the most appropriate means of countering any potential attacks is required. A complete implementation of these techniques needs to be done and the results evaluated using real world data.

Appendix A

Jamming Without Exact Knowledge of Receiver Location

It is clear from the results shown in Chapter 3 that to jam an FM PR, the important parameter is the $\mathrm{JSR_E}$ in the surveillance channel. J is defined as the direct path jamming power at the receiver and S_E is defined as the target echo power at the receiver. The $\mathrm{JSR_E}$ is therefore defined as

$$J = \frac{P_J G_J G_{Rj} \lambda^2}{(4\pi)^2 R_J^2} \tag{A.1}$$

$J =$ Direct path jammer signal power at surveillance receiver.
$P_J =$ Jammer transmit power.
$G_J =$ Jammer antenna gain.
$G_{Rj} =$ Receiver antenna gain in direction of jammer.
$R_J =$ User defined 'Jammer Radius' over which jammer is intended to be effective.
$\lambda =$ Wavelength.

$$S_E = \frac{P_T G_T G_{Rt} \lambda^2 \sigma}{(4\pi)^3 R_{Tx}^2 R_{Rx}^2} \tag{A.2}$$

$S_E = $ Target echo power at the surveillance receiver.

$P_T = $ Transmitter of opportunity transmit power.

$G_T = $ Transmit antenna gain in direction of target.

$G_{Rt} = $ Receive antenna gain in direction of target.

$R_{Tx} = $ Transmitter to target distance.

$R_{Rx} = $ Target to receiver distance.

$\sigma = $ Bistatic RCS of potential target.

$\lambda = $ Wavelength.

$$
\begin{aligned}
J/S_E &= \left[\frac{P_J G_J G_{Rj} \lambda^2}{(4\pi)^2 R_J^2}\right] \div \left[\frac{P_T G_T G_{Rt} \lambda^2 \sigma}{(4\pi)^3 R_{Tx}^2 R_{Rx}^2}\right] \\
&= \left[\frac{ERP_J G_{Rj} 4\pi}{ERP_T G_{Rt} \sigma}\right]\left[\frac{R_{Tx}^2 R_{Rx}^2}{R_J^2}\right]
\end{aligned}
\tag{A.3}
$$

Rearranging (A.3) such that the required jammer ERP ($ERP_J = P_J G_J$) is the subject gives:

$$
ERP_J = \left[\frac{J}{S_E}\right]\left[\frac{ERP_T G_{Rt} \sigma}{G_{Rj} 4\pi}\right]\left[\frac{R_J^2}{R_{Tx}^2 R_{Rx}^2}\right]
\tag{A.4}
$$

The required ERP_J can be minimised if the exact location of the receiver is known. This would allow the operator to select a highly directional antenna and place it within the main beam of the surveillance receiver while minimising the distance between the jammer and the receiver.

In typical deployment scenarios, the information available to the ECM operator is less than ideal and it is therefore important to investigate a more general solution to the problem. Analysing (A.4) three different performance evaluation scenarios are defined. The first assumes no knowledge of the receiver other than the fact that it is surveying a certain volume of space. A jammer can then be deployed such that a target of interest flies within what is defined in Section A.1 as the 'jammer radius', i.e. the distance from the jammer as defined by the operator. The second scenario assumes that a receiver to be jammed is within the jammer radius throughout the flight duration, i.e. the jammer is placed in such a position that any receiver within the jammer radius will be jammed. The third scenario relies on self-protection or escort jamming where the target itself carries the jammer. Each scenario is discussed in detail in Sections A.1 to A.3.

This work focuses on the effectiveness of jamming on an FM based PR. Effectiveness therefore references performance from the jammer perspective. This implies that unless explicitly stated, 'worst case' is defined from the jammers perspective.

A.1 Target Within Jammer Radius

The first scenario described, requires that the target to be masked or jammed lies within the jammer radius, represented by R_J. The jammer radius is defined as a distance away from the jammer that is selected by the jammer operator. In this case, the jammer radius can be dynamic in that it can change depending on where the target is in the scene. Assuming the operator knows the flight path of the target or targets, this radius can be adjusted accordingly as the target moves relative to the jammer location. The concept is illustrated in Figure A.1.

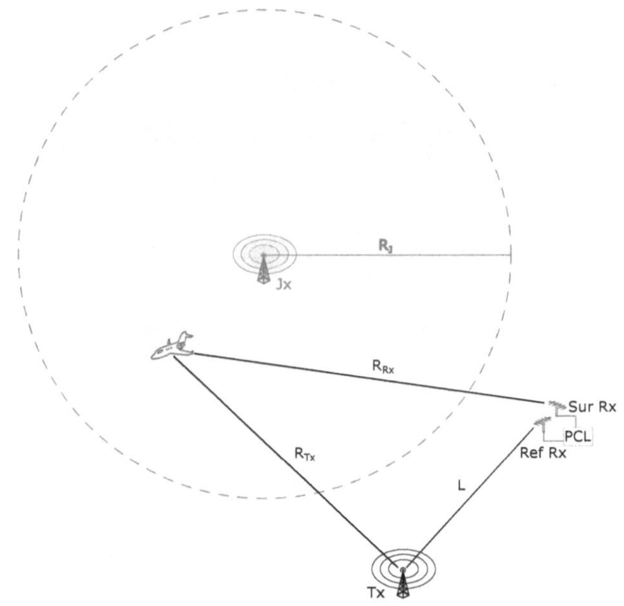

Figure A.1: Scenario with target within the jammer radius and the receiver in an unknown location. The location of the receiver is unknown. The potential transmitter locations are known to the operator via official documentation.

To ensure the effective jamming of any potential receiver, the JSR_E must be maintained at the desired level. This level can be calculated by referencing the position of the target to the transmitter and jammer. The jammer power at the target position along the flight path can be calculated using the one way path loss equation (A.1). Similarly, the incident power on the target from the transmitter can be calculated and the resultant echo power reflected off the target can be calculated. Regardless of where the receiver lies, the distance from the target to the receiver, denoted by R_{Rx}, will always be less than or equal to the altitude of the aircraft, assuming a ground based PR. This implies that, provided the jammer is using an antenna with a uniform azimuth radiation pattern (i.e. a dipole antenna), any path loss that the jamming signal receives will be less than or equal to the path loss incurred by the target echo, with equal attenuation being when the receiver is on the same line as the target-jammer and worse case (from a jamming perspective) being when the target-receiver distance, R_{Rx}, is shorter than the jammer-receiver distance.

This approach can then be used to determine the parameters needed to ensure that the required JSR_E is achieved for any ground based PR, given that the target remains within the jammer radius. The required ERP_J from (A.4) is therefore modified as follows:

$$J_{incident} = \frac{P_J G_J G_{Rj} \lambda^2}{(4\pi)^2 R_{J_{incident}}^2} \tag{A.5}$$

$$S_E = \frac{P_T G_T G_{Rt} \lambda^2 \sigma}{(4\pi)^3 R_{Rx}^2 R_{Tx}^2} \tag{A.6}$$

where $J_{incident}$ is the jammer power at the target location and S_E is the echo power at the target location. The distance from the target to the receiver is unknown to the jammer operator. To account for this unknown, the distance between the jammer and the receiver can be modelled as:

$$\begin{aligned} R_{J_{incident}}^2 &= (R_J + R_{Rx})^2 \\ &= R_J^2 + 2 R_J R_{Rx} + R_{Rx}^2 \end{aligned} \tag{A.7}$$

with R_J being the jammer radius to the target and R_{Rx} being the distance from the target to the receiver. The target-to-receiver distance can be broken down into the ground distance and the target altitude above ground such that:

$$R_{R_x} = \sqrt{R_{R_{gx}}^2 + R_H^2} \tag{A.8}$$

where R_{Rx} represents the target-to-receiver ground distance and R_H represents the targets altitude above the ground. This is a worst case scenario whereby the jammer operator assumes that the receiver is positioned in line with the jammer-to-target line, outside the jammer radius. In the event that the receiver is situated at any location other than along the jammer-to-target line, the distance from the jammer-to-receiver could be considerably smaller. The ERP_J required to produce a JSR_E at the receiver is therefore:

$$\frac{J}{S_E} = \left[\frac{P_J G_J G_{Rj}\lambda^2}{(4\pi)^2(R_J^2 + 2R_J\sqrt{R_{R_{gx}}^2 + R_H^2} + R_{R_{gx}}^2 + R_H^2)}\right]\left[\frac{(4\pi)^3 sqrt(R_{R_{gx}}^2 + R_H^2)^2 R_{Tx}^2}{P_T G_T G_{Rt}\lambda^2\sigma}\right]$$

$$(A.9)$$

$$\therefore ERP_J = \left[\frac{J}{S_E}\right]\left[\frac{ERP_T G_{Rt}\sigma}{G_{Rj}(4\pi)}\right]\left[\frac{R_J^2 + 2R_J\sqrt{R_{R_{gx}}^2 + R_H^2} + R_{R_{gx}}^2 + R_H^2}{(R_{R_{gx}}^2 + R_H^2)R_{Tx}^2}\right] \qquad (A.10)$$

where JSR_E is the required jammer to signal echo level at the receiver for effective jamming and G_{Rt} and G_{Rj} are the receive gains in the direction of the target and jammer respectively with σ being the target RCS.

Evaluating the last part of (A.10), it is clear that the target-receiver distance plays a major role. A worst case scenario would be when the target is flying directly above the receiver, provided that it is still within the mainbeam of the receivers surveillance antenna. We can therefore say that the worst case value for R_{Rgx} is zero resulting in an ERP_J of:

$$ERP_J = \left[\frac{J}{S_E}\right]\left[\frac{ERP_T G_{Rt}\sigma}{G_{Rj}(4\pi)}\right]\left[\frac{R_J^2 + 2R_J R_H + R_H^2}{R_H^2 R_{Tx}^2}\right] \qquad (A.11)$$

A.2 Receiver Within Jammer Radius

The scenario presented in Section A.1 assumes no prior knowledge of the receiver and therefore requires a relatively large amount of jamming power to be sure of jamming any potential receivers. The jamming power can be significantly reduced if the assumption can be made that the receiver lies within the so-called 'jammer radius'. The jammer radius is therefore defined in this context as the distance away from the jammer that guarantees the jamming of any receiver within the radius. Once again we define the jammer-to-signal echo ratio using (A.1) and (A.2) that results in a required ERP_J of:

$$ERP_J = \left[\frac{J}{S_E}\right]\left[\frac{ERP_T G_{Rt}\sigma}{G_{Rj}4\pi}\right]\left[\frac{R_J^2}{R_{Tx}^2 R_{Rx}^2}\right] \qquad (A.12)$$

The path distance between a potential transmitter and target can be calculated while the path distance from the target to the receiver is unknown, other than it must be within the jammer radius, R_J. This is complicated further in that the target could lie either inside or outside the jammer radius.

A. Target lies outside the jammer radius

An example scenario where the target lies on the outside of the jammer radius while the receiver lies within it is shown in Figure A.2.

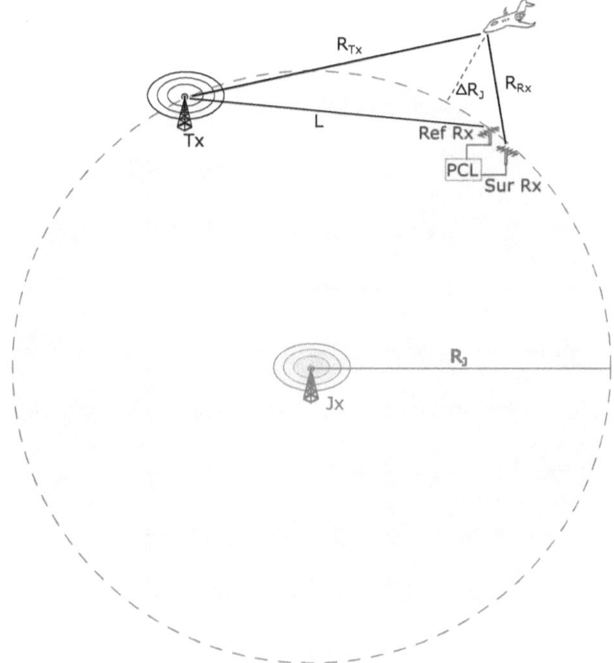

Figure A.2: Scenario with target outside jammer radius and receiver within jammer radius.

The blue dashed line represents the ground distance between the target and the edge of the jammer radius, denoted by ΔR_J. Breaking down (A.12), the jammer radius, R_J is set by the operator and R_{Tx} is determined based on the flight geometry. The value for R_{Rx} depends on whether the target remains inside or outside the jammer radius. For the target outside the jammer radius, the minimum possible value of R_{Rx} is determined by the altitude of the target, R_H, as well as its distance from the edge of the jammer radius, ΔR_J, such that:

$$R_{Rx} = \sqrt{R_H^2 + \Delta R_J^2} \qquad (A.13)$$

This results in a required ERP_J of

$$ERP_J = \left[\frac{J}{S_E}\right]\left[\frac{ERP_T G_{Rt}\sigma}{G_{Rj}4\pi}\right]\left[\frac{R_J^2}{R_{Tx}^2(R_H^2 + \Delta R_J^2)}\right] \qquad (A.14)$$

B. Target lies within the jammer radius

An example scenario where the target lies within the jammer radius is shown in Figure A.3.

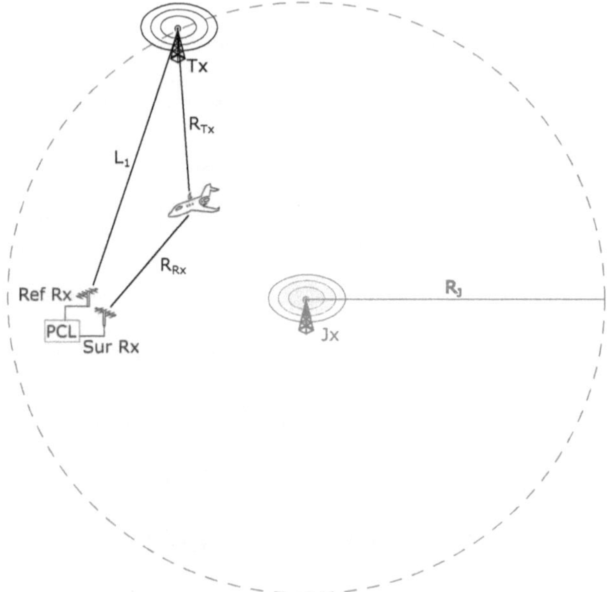

Figure A.3: Scenario with target inside jammer radius and receiver within jammer radius.

With both the receiver and target being within the jammer radius, the worst case value of R_{Rx} is achieved when the target is closest to the receiver, i.e. above it. R_{Rx} becomes the altitude of the target, R_H. The resulting worst case ERP_J is then given by:

$$ERP_J = \left[\frac{J}{S_E}\right]\left[\frac{ERP_T G_{Rt}\sigma}{G_{Rj}4\pi}\right]\left[\frac{R_J^2}{R_{Tx}^2 R_H^2}\right] \tag{A.15}$$

Combining the results of (A.14) and (A.15) results in an ERP range of

$$\left[\frac{J}{S_E}\right]\left[\frac{ERP_T G_{Rt}\sigma}{G_{Rj}4\pi}\right]\left[\frac{R_J^2}{R_{Tx}^2 (R_H^2 + \Delta R_J^2)}\right] \leq ERP_J \leq \left[\frac{J}{S_E}\right]\left[\frac{ERP_T G_{Rt}\sigma}{G_{Rj}4\pi}\right]\left[\frac{R_J^2}{R_{Tx}^2 R_H^2}\right] \tag{A.16}$$

A.3 Self Protection Jamming and Escort Jamming

The scenarios presented in Sections A.1 and A.2 represented forms of stand-off jamming where a separate jammer is used to jam a receiver or mask a target. This approach can lead to considerably more power being required if the receiver location is unknown.

Self protection jamming or escort jamming can therefore be used as an effective means of jamming an FM based PR when the location of the receiver is unknown. This is demonstrated by illustrating the relationship between the jammers required ERP and the incident power on the target.

The target echo power at the receiver is given by

$$S_E = \frac{P_T G_T G_{Rt}\lambda^2\sigma}{(4\pi)^3 R_{Tx}^2 R_{Rx}^2} \tag{A.17}$$

Using an isotropic antenna, the jammer power received at the receiver is

$$J = \frac{P_J G_J G_{Rj}\lambda^2}{(4\pi)^2 R_J^2} \tag{A.18}$$

Since the receiver gain is the same for the target echo, G_{Rt}, and the jammer signal, G_{Rj}, and the target-receiver distance, R_{Rx}, the same as the jammer-receiver distance, R_J, the resulting jammer ERP becomes

$$ERP_J = \left[\frac{J}{S_E}\right]\left[\frac{ERP_T\sigma}{4\pi R_{Tx}^2}\right] \tag{A.19}$$

A.4 Section Summary

It is clear that an ECM operator requires some form of prior knowledge of the system to be jammed. The first is the desired jammer to signal ratio. The second is the potential scenario specifications such as the target to be masked as well as the transmitter band that is being used. This information is readily available from public institutions such as the countries regulatory bodies website etc. The final piece of prior knowledge required for successful jamming is the targets location relative to the transmitter. This can be achieved through careful planning.

There are four different approaches that can be used to ensure jamming of any potential target, these include:

1. **Receiver location known**

$$ERP_J = \left[\frac{J}{S_E}\right]\left[\frac{ERP_T G_{Rt}\sigma}{G_{Rj}4\pi}\right]\left[\frac{R_J^2}{R_{Tx}^2 R_{Rx}^2}\right] \tag{A.20}$$

- Complete knowledge of receiver location is required.
- Most effective solution requiring the least power to jam.

2. **Target lies within jammer radius**

$$ERP_J = \left[\frac{J}{S_E}\right]\left[\frac{ERP_T G_{Rt}\sigma}{G_{Rj}4\pi}\right]\left[\frac{R_J^2 + 2R_J R_H + R_H^2}{R_{Tx}^2 R_H^2}\right] \tag{A.21}$$

- No knowledge of possible receiver location.
- Requires high power levels to achieve desired J/S_E.

3. **Receiver lies within jammer radius**

- Two possible options:

  - Target lies outside jammer radius R_J.

$$ERP_J = \left[\frac{J}{S_E}\right]\left[\frac{ERP_T G_{Rt}\sigma}{G_{Rj}4\pi}\right]\left[\frac{R_J^2}{R_{Tx}^2 (R_H^2 + \Delta R_J^2)}\right] \tag{A.22}$$

  - Target lies within jammer radius R_J.

$$ERP_J = \left[\frac{J}{S_E}\right]\left[\frac{ERP_T G_{Rt}\sigma}{G_{Rj}4\pi}\right]\left[\frac{R_J^2}{R_{Tx}^2 R_H^2}\right] \tag{A.23}$$

  - Knowledge of potential receiver locations.

- Provides significant reduction in required power to achieve desired JSR_E.

4. **Self protection jamming**

$$ERP_J = \left[\frac{J}{S_E}\right]\left[\frac{ERP_T\sigma}{4\pi R_{Tx}^2}\right] \tag{A.24}$$

- Most effective option when receiver location is unknown.

It is clear that the performance of an FM based PR is highly dependant on a variety of factors, including antenna location, DSI cancellation as well as any potential interference or jamming. Using the results shown in Chapters 3, we demonstrate the feasibility of jamming an FM based PR over a particular volume.

A.5 Standoff Jamming

A.5.1 Known Rx Location

From (A.20) we can determine the minimum and maximum ERP required to cause the minimum desired reduction in performance of the target PR given a set of scenario parameters. As an example of a potential scenario, we investigate a flight path and geometry applicable to the Western Cape area of South Africa as illustrated by Figure A.6.

Figure A.4: Google maps overhead view of FM transmitter locations throughout the Western Cape area. The ERP of each transmitter is shown below the transmitter name.

It has already been demonstrated that if the receiver location is known, low jamming power can be used to effectively mask the target of interest. Assuming a complete scenario geometry such as the one shown in Figure A.5, (A.20) can be used to determine the required jammer ERP.

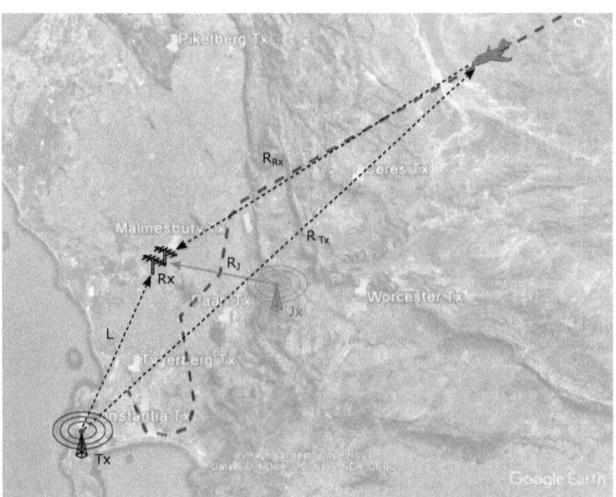

Figure A.5: Scenario 1 - Known receiver location where target flies a typical flight path into Cape Town International airport. This scenario is similar to the one that has been simulated with the receiver and transmitter locations known.

Along the flight path of the target, the smallest value for the bistatic range will be achieved when the target comes in to land (ignoring the fact that the target would inherently be masked due to the high powered direct signal and possible nulls in the beam pattern as a result.). The parameters used in this scenario are shown in Table A.1

Table A.1: Parameters for highest echo return along flight path shown in Figure A.5

Parameter	Value
ERP_T	10 kW
R_{Tx}	49 km
R_{Rx}	30 km
R_J	34 km
L	71 km

The desired JSR_E can be selected from Table 4.1. Using a medium bandwidth FM jamming ($\beta = 2$) waveform and a 50% reduction in system performance, assuming a co-located receiver system, the required jammer ERP can then be determined as

$$ERP_J = \left[\frac{J}{S_E}\right] \left[\frac{ERP_T G_{Rt} \sigma}{G_{Rj} 4\pi}\right] \left[\frac{R_J^2}{R_{Tx}^2 R_{Rx}^2}\right]$$

$$= \left[46 \text{ dB}\right] \left[\frac{10\,000 \text{ kW} \times 200 \text{ m}^2}{4\pi}\right] \left[\frac{(34\,000 \text{ m})^2}{(49\,000 \text{ m})^2 (34\,000 \text{ m})^2}\right] \tag{A.25}$$

$$= 46 \text{ dB} + (-41.8 \text{ dB})$$

$$= 5.2 \text{ dB}$$

$$= 3.3 \text{ W}$$

The result from equation (A.25) coincides with the results shown in Chapter 3 where the receiver can be jammed using a very low amount of jamming power (3.3 W in this case).

A.5.2 Unknown Rx Location

Target Within Jammer Radius

The scenario whereby the exact location of the receiver is known is a very specific example in that this would not necessarily be the case in practice. The scenario can be generalised for an unknown receiver location. Figure A.6 illustrates an example scenario similar to that shown in A.5 but with an unknown receiver location.

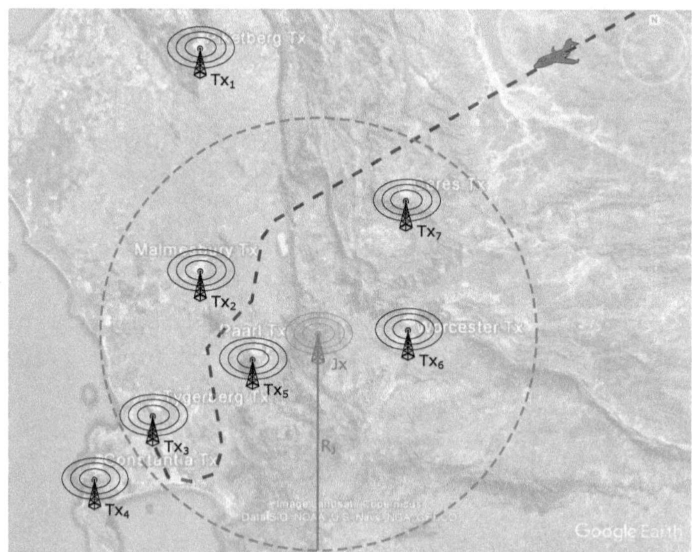

Figure A.6: Possible flight path into the Western Cape area from Johannesburg to be jammed. Each transmitter is highlighted while a typical flight path from Johannesburg to Cape Town International airport is shown in blue. A jammer is placed with a 'jammer radius', R_j, such that the entire flight path of interest is covered.

To determine the required jammer ERP, ERP_J, we need to evaluate the particular geometry of each of the potential transmitters in the vicinity. Assuming no additional knowledge of any potential PR, other than that it operates over the FM band, equation (A.21) will need to be evaluated for the given scenario as shown

$$
\begin{aligned}
ERP_J &= \left[\frac{J}{S_E}\right]\left[\frac{ERP_T G_{Rt}\sigma}{G_{Rj}(4\pi)}\right]\left[\frac{R_J^2 + 2R_J R_H + R_H^2}{R_H^2 R_{Tx}^2}\right] \\
&= \left[\frac{J}{S_E}\right]\left[\frac{G_{Rt}\sigma(R_J^2 + 2R_J R_H + R_H^2)}{G_{Rj}(4\pi)R_H^2}\right]\left[\frac{ERP_T}{R_{Tx_{min}}^2}\right]
\end{aligned}
\tag{A.26}
$$

where

$$
\left[\frac{J}{S_E}\right] = \text{Read off Table 4.1.}
$$

$$
\left[\frac{G_{Rt}\sigma(R_J^2+2R_J R_H+R_H^2)}{G_{Rj}(4\pi)R_H^2}\right] = \text{Determined based on scenario requirements in Figure A.6.}
$$

$$
\left[\frac{ERP_T}{R_{Tx_{min}}^2}\right] = \text{Determined by careful evaluation of Figure A.6 as shown in Table}
$$

Table A.2 summarises the transmitter parameters required to evaluate (A.26). The higher the resultant value, the worse the performance of the jammer and therefore the highest value is used when solving (A.26).

Table A.2: Transmitter ERPs and minimum distance to flight path for each of the transmitters shown in Figure A.4.

Transmitter	ERP_T [kW]	$R_{Tx}min$ [km]	$\frac{ERP_T}{R_{Tx}^2 min}$
Piketberg	10	60	2.78×10^{-6}
Malmesbury	10	21	2.27×10^{-5}
Tygerberg	1.3	5	5.2×10^{-5}
Constantia	10	21	2.27×10^{-5}
Paarl	0.3	9	3.7×10^{-6}
Worcester	0.1	48	4.34×10^{-8}
Ceres	20	16	7.81×10^{-5}

The highest value is achieved using the Ceres site with 7.81×10^{-5}. The remaining parameters are shown in Table A.3.

Table A.3: Worst case system parameters for Figure A.6.

Parameter	Value
ERP_T	20 kW
R_T	16 km
R_J	40 km
R_H	10 km
G_{Rt}	G_{Rj}
σ	200 m^2

The required jammer power can then be calculated by choosing a suitable waveform and JSR$_E$ level from Table 4.1. In this example, medium bandwidth FM jamming ($\beta = 2$) with a 46 dB JSR$_E$ is used to cause a 50% reduction in overall system performance. Assuming the target is flying at cruising altitudes of 10 000 m Above Mean Sea Level (AMSL), the required jammer ERP is calculated as

$$
\begin{aligned}
ERP_J &= \left[\frac{J}{S_E}\right]\left[\frac{ERP_T G_{Rt}\sigma}{G_{Rj}(4\pi)}\right]\left[\frac{R_J^2 + 2R_J R_H + R_H^2}{R_H^2 R_{Tx}^2}\right] \\
&= \left[46 \text{ dB}\right]\left[\frac{20\,000 \text{ W} \cdot 200 \text{ m}^2}{4\pi}\right]\left[\frac{(40\,000 \text{ m})^2 + 2(40\,000 \text{ m})(10\,000 \text{ m}) + (10\,000 \text{ m})^2}{(10\,000 \text{ m})^2(16\,000 \text{ m})^2}\right] \\
&= 46 \text{ dB} - 15 \text{ dB} \\
&= 1.26 \text{ kW}
\end{aligned}
$$

$$(A.27)$$

An ERP$_J$ of 1.26 kW is significant orders of magnitude more than the 3.3 W shown previously, however, this applies to the worst case scenario whereby the receiver is situated on the absolute edge of the jammer radius. Making a few small adjustments to the flight path could lead to a dramatic reduction in overall required power.

Receiver Within Jammer Radius

By making the assumption that the PR that is to be jammed is situated within the jammer radius, the total jamming power required to ensure jamming can be reduced considerably. Using the same scenario described in Section A.5.2 with one of the differences being the receiver lying within the jammer radius, the required ERP can be determined. The lowest ERP requirement occurs when the target is furthest away from the receiver, i.e. the target is furthest away from the jammer radius. In this scenario, lets assume the furthest the target is away from the jammer radius is 20 km as shown in Figure A.6. The desired JSR$_E$ can be selected from Table 4.1. Using the same jamming medium bandwidth FM jamming ($\beta = 2$) waveform and a 50% reduction in system performance, the minimum required jammer ERP can then be determined as

$$
\begin{aligned}
ERP_J &= \left[\frac{J}{S_E}\right]\left[\frac{ERP_T G_{Rt}\sigma}{G_{Rj}4\pi}\right]\left[\frac{R_J^2}{R_{Tx}^2(R_H^2 + \Delta R_J^2)}\right] \\
&= \left[46\ \text{dB}\right]\left[\frac{20\,000\ \text{W} \cdot 200\ \text{m}^2}{4\pi}\right]\left[\frac{(40\,000\ \text{m})^2}{(16\,000\ \text{m})^2(10\,000^2\ \text{m} + 20\,000^2\ \text{m})}\right] \\
&= 46\ \text{dB} - 24\ \text{dB} \\
&= 22\ \text{dB} \\
&= 158.5\ \text{W}
\end{aligned}
$$

(A.28)

As the target approaches its destination and enters the jammer radius, the jamming power requirement changes to

$$
\begin{aligned}
ERP_J &= \left[\frac{J}{S_E}\right]\left[\frac{ERP_T G_{Rt}\sigma}{G_{Rj}4\pi}\right]\left[\frac{R_J^2}{R_{Tx}^2 R_H^2}\right] \\
&= \left[46\ \text{dB}\right]\left[\frac{20\,000\ \text{W} \cdot 200\ \text{m}^2}{4\pi}\right]\left[\frac{(40\,000\ \text{m})^2}{(16\,000\ \text{m})^2(10\,000\ \text{m})^2}\right] \\
&= 46\ \text{dB} - 17\ \text{dB} \\
&= 794\ \text{W}
\end{aligned}
$$

(A.29)

It is clear that knowledge of the existence of a receiver in a particular area allows the jammer ERP to be reduced considerably. Further knowledge of a receivers location would allow for higher gain antennas to be used that offer high directionality and therefore result in reduced power requirements.

One method that can be used to reduce the jamming power required is to use multiple receivers scattered around the flight path. This will allow more area to be covered while maintaining a lower power level for each jamming site.

An example of improving the jamming performance with additional knowledge of the potential receiver is illustrated in Figure A.7. If it is known that a receiver lies within a particular region and the number of possible transmitters is narrowed down due to geometry limitations such as the mountain range shown in Figure A.7, the required jammer power level can be reduced.

Figure A.7: Refined jammer scenario with receiver located within jammer radius. A receiver is assumed to be at either Malmesbury or at UCT (as has been the case with practical field trials).

With receiver sites located somewhere within the jammer radius, specifically at the UCT or Malmesbury sites as has been the case with previous field trials, the transmitters that can be used are narrowed down with the Ceres and Worcester transmitters falling away. Looking at Table A.2, the highest value given the new geometry is achieved with the Tygerberg transmitter. The issue with the Tygerberg transmitter as a source is that unless used as a forward scatter radar, a target flying close to the transmitter would be masked by the direct signal interference. This would severely limit the performance of the radar should the target travel near the transmitter.

The new system parameters are therefore summarised as

Table A.4: Worst case system parameters for Figure A.7.

Parameter	Value
ERP_T	1.3 kW
R_T	5 km
R_J	30 km
G_{Rt}	G_{Rj}
σ	200 m^2

Using the same medium bandwidth jamming waveform at 46 dB $\mathrm{JSR_E}$ as previously, the required worst case $\mathrm{ERP_J}$ can therefore be calculated as

$$
\begin{aligned}
ERP_J &= \left[\frac{J}{S_E}\right]\left[\frac{ERP_T G_{Rt}\sigma}{G_{Rj}4\pi}\right]\left[\frac{R_J^2}{R_{Tx}^2 R_H^2}\right] \\
&= \left[46\text{ dB}\right]\left[\frac{1\,300\text{ W}\cdot 200\text{ m}^2}{4\pi}\right]\left[\frac{(30\,000\text{ m})^2}{(5\,000\text{ m})^2(10\,000\text{ m})^2}\right] \\
&= 46\text{ dB} - 21.3\text{ dB} \\
&= 295\text{ W}
\end{aligned}
\tag{A.30}
$$

This results in a maximum $\mathrm{ERP_J}$ reduction from 794 W to 295 W, a 4.3 dB reduction in required $\mathrm{ERP_J}$.

A.6 Self Protection or Escort Jamming

Self protection jamming and escort jamming would be highly effective against an FM based PR as it would be trivial to detect the transmitter power incident on the target at any given point and simply transmit a Gaussian noise waveform at the appropriate power level in order to achieve the desired $\mathrm{JSR_E}$ with as little transmit power as possible.

Assuming the same scenario as described in Section A.5.1, the maximum incident transmit power from any transmitter is shown in Table A.5.

Table A.5: Maximum incident power reflected off the target along the flight path shown in Figure A.6 for each transmitter.

Transmitter	ERP_T [kW]	$R_{Tx}min$ [km]	$P_{target} = \frac{ERP\sigma}{4\pi R_{Tx}^2}$ [dB]
Piketberg	10	60	-43.5
Malmesbury	10	21	-34.4
Tygerberg	1.3	5	-30.8
Constantia	10	21	-34.4
Paarl	0.3	9	-42.3
Worcester	0.1	48	-61.6
Ceres	20	16	-29

The maximum incident power that is radiated off the target occurs when the target is close to the Ceres transmitter at -29 dB. In order for any potential FM based PR to be jammed using AWGN, the self protection jammer would then need to transmit broadband noise across the FM spectrum with an ERP determined by Table 4.1.

To guarantee a reduction in detection performance of at least 50% using a medium bandwidth FM jamming ($\beta = 2$) waveform, the JSR_E is required to be at least 46 dB. The worst case jammer ERP would therefore need to be 17 dB i.e. 50 W. It is important to note that a target applying self protection jamming does not want to radiate 50 W of RF as it would be problematic from an ES point of view. E.g. a noise jammer could be located relatively simply through multilateration that could be achieved by a network of spatially separated PR sensors.

This could be significantly reduced if additional assumptions were made by the jammer operator. Given the nature of passive radar, the surveillance antenna would need to have its antenna beam directed away from the reference transmitter due to the high direct signal interference impact on receiver sensitivity. As a result, it can be said that a target flying within a particular radius of the transmitter would be inherently masked.

A.7 Conclusions

The exact performance of the jammer is closely tied to the geometry of the system relative to the jammer. It has been shown that knowledge of the receiver location plays a major role in determining the required jammer power in order to effectively jam the radar receiver. In an ideal sense where the locations of both the receiver and transmitter being used are known, the receiver can be jammed with using very low power levels and a directional antenna. In the case where the receiver location is unknown, an omni directional antenna or at the very least, an antenna with a wide beamwidth will be required, this therefore increases the amount of jamming power required.

The most effective and reliable form of jamming of an FM based passive radar when the receiver location is unknown is self protection or escort jamming. This is due to the fact that in order to jam the radar using Gaussian noise-like waveforms, the jammer power needs to be greater than the target echo power. The target echo power can be determined by measuring the incident energy on the target from any particular transmitter and could then transmit a jamming signal proportional to the received energy. This would guarantee the lowest amount of power used to mask the target from a receiver in an unknown location.

Appendix B

Direct Signal Suppression Algorithms

This Appendix provides a brief overview of the two cancellation algorithms utilised in this work. As discussed in Chapter 3, the two techniques used in this work are ECA and CGLS.

To remove the DSI from the surveillance signal, S_{surv}, we need to subtract the interfering signal, S_{dsi}, leaving (ideally) only the target echos. To produce an estimate of S_{dsi}, the matrix A can be constructed by building a matrix from the recorded reference signal and zero padding according to each bistatic range bin that you want to suppress the DSI. We then require an estimate of the scaling coefficients, x, which are then used to solve the equation $S_{dsi} = Ax$.

As the number of columns in A are made up of the number of delays and the rows represent the number of samples, the matrix A is typically not square and is therefore not invertable. This means that the equation $S_{dsi} = Ax$ is not directly solvable. The solution is therefore to minimise using a least squares approach.

B.1   Least Squares Regressive Theory

As described in [11], [72] and [153], in order to understand the ECA process, it is critical to understand the underlying principle of which it is based off. Normal regression is the process of fitting a general linear additive model, described as:

$$b = a_1 f_1(x) + a_2 f_2(x) + ... + a_n f_n(x) \tag{B.1}$$

which can be represented using matrix notation as:

$$b = Ax \tag{B.2}$$

Here, b is a vector of observed values while x is a vector of filter coefficients. A is a design matrix where each column consists of a linear predictor function $f(x)$:

$$b = \begin{bmatrix} b_1 & b_2 & \cdots & b_n \end{bmatrix}^T$$

$$x = \begin{bmatrix} a_1 & a_2 & \cdots & a_n \end{bmatrix}^T$$

$$A = \begin{bmatrix} f_1(x) & f_2(x) & \cdots & f_n(x) \end{bmatrix}$$

The error term, $e = Ax - b$ is simply the difference between the observed values b and the predicted values Ax. The least squares solution for calculating the filter coefficients x is that which minimises the error term:

$$min| \sum_{i=1}^{n} (Ax - b)^2 | \tag{B.3}$$

By taking the derivative of the error term, e, with respect to the filter coefficients x, it is shown that the values of x which minimise e are given by:

$$\begin{aligned} x &= A^* b \\ &= [(A^T A)^{-1} A^T] b \end{aligned} \tag{B.4}$$

where A^* denotes matrix pseudo-inversion. Having calculated the filter coefficients, the residual term between the observed and predicted values can then be expressed as:

$$\begin{aligned} e &= [(A^T A)^{-1} A^T] b - b \\ &= (A(A^T A)^{-1} A^T - I_n) b \\ &= Pb \end{aligned} \tag{B.5}$$

The projection matrix, P, is therefore represented as:

$$P = A(A^T A)^{-1} A^T - I_n \tag{B.6}$$

which has the effect of mapping b onto the vector subspace spanned by columns of A.

It is the formation of the projection matrix P that is ultimately exploited for DSI suppression. Specifically, equation B.3 can be modified to the mathematical equivalent of:

$$min | \sum_{i=1}^{n} (b - Ax)^2 | \tag{B.7}$$

This results in the projection matrix:

$$P' = I_n - A(A^T A)^{-1} A^T \tag{B.8}$$

which, unlike the projection matrix in equation B.6, P, which projects towards b, the projection matrix in equation B.8, P', projects away from b, essentially removing the DSI.

B.2   Extensive Cancellation Algorithm

As discussed, the ECA algorithm aims to solve the equation:

$$min | S_{surv} - Ax |^2 \tag{B.9}$$

In order to suppress DSI in the surveillance channel, S_{surv}, the clutter subspace matrix, A, is constructed from frequency shifted, delayed and scaled replicas of the reference signal, S_{ref}. Construction of the clutter matrix, A, is achieved by creating the zero-Doppler profile:

$$A_{zero-Doppler} = \begin{bmatrix} S_{Ref}[0] & 0 & ... & 0 \\ S_{Ref}[1] & S_{Ref}[0] & ... & 0 \\ ... & ... & ... & ... \\ S_{Ref}[N_s - 1] & S_{Ref}[N_s - 2] & ... & S_{Ref}[N_s - (K - 1)] \end{bmatrix}$$

where K is the number of range bins to be suppressed along the zero-Doppler line. To account for non-stationary clutter, the final A matrix is constructed by applying p

frequency shifts to each column of the $A_{zero-Doppler}$ matrix.

$$A = \begin{bmatrix} \Delta_{-p}A_{zero-Doppler} & \cdots & A_{zero-Doppler} & \cdots & \Delta_p A_{zero-Doppler} \end{bmatrix}$$

where p is the desired Doppler shift. For $-p < i < p$, the diagonal matrix Δ_i is responsible for applying the relevant frequency shift and is constructed as:

$$\Delta_i \begin{bmatrix} 1 & 0 & \cdots & 0 \\ 0 & e^{j2\pi*i} & \cdots & 0 \\ \cdots & \cdots & \cdots & \cdots \\ 0 & 0 & \cdots & e^{j2\pi*i(N_s-1)} \end{bmatrix}$$

With the clutter subspace matrix constructed, the adaptive filter coefficients, x, can then be calculated as described in equation B.4 in Section B.1 as:

$$\alpha = (A^H A)^{-1} A^H S_{surv} \tag{B.10}$$

The resultant cancelled surveillance signal is therefore represented as:

$$S_{cancelled_E CA} = S_{surv} - Ax \tag{B.11}$$

While ECA provides excellent DSI cancellation performance, it is computationally expensive due to its "one-shot" nature. An alternative to ECA is the iterative CGLS process as described in Section B.3.

B.3　Conjugate Gradient Least Squares

As shown in [72], the CGLS algorithm is an extension of the conjugate gradient technique whereby with each iteration, the gradient is chosen to be conjugate to the previous gradient. This can then be used to minimise $S_{dsi} = Ax$ using the least squares approach described in Section B.1.

The CGLS algorithm achieves this by iteratively tending towards a minimum value. Once a satisfactory residual is achieved, the processing can be stopped. Alternatively, the algorithm can be run for a fixed number of iterations, regardless of whether a minimum has been achieved or not. The advantage of this approach over ECA is that it results in fixed execution times and a fixed memory footprint as there is no matrix inversion requirement. This makes CGLS highly suitable for real-time operation.

In typical static clutter environments, the DSI remains fairly constant over time and therefore a fixed number of iterations, once converged, can be used to adequately remove the DSI from the surveillance channel as demonstrated in [11]. An excellent, in depth technical report on CGLS is presented by Shewchuk in [154] for further information.

Appendix C

DVB-T2 System Parameters

Table C.1: DVB-T2 scattered pilot pattern parameters.

Scattered Pilot (SP) Pattern	Separation of pilot-bearing carriers (P_s)	No. of symbols forming one SP sequence (C_s)	Amplitude (A_{SP})	Equivalent Boost [dB]
PP1	3	4	4/3	2.5
PP2	6	2	4/3	2.5
PP3	6	4	7/4	4.9
PP4	12	2	7/4	4.9
PP5	12	4	7/3	7.4
PP6	24	2	7/3	7.4
PP7	24	4	7/3	7.4
PP8	6	16	7/3	7.4

Table C.2: DVB-T2 continual and P2 pilot parameters for different FFT sizes.

Pilot Amplitude	1K	2K	4K	8K	16K	32K
			FFT Size			
A_{CP}	4/3	4/3	$(4\sqrt{2})/3$	8/3	8/3	8/3
A_{P2}	$\sqrt{31}/5$	$\sqrt{31}/5$	$\sqrt{31}/5$	$\sqrt{31}/5$	$\sqrt{31}/5$	$\sqrt{37}/5$

Table C.3: Summary of active and pilot carriers for different DVB-T2 FFT sizes.

		1K	2K	4K	8K N	8K E	16K N	16K E	32K N	32K E
Total No. Active Carriers		853	1 705	3 409	6 817	6 913	13 633	13 921	27 265	27 841
PP1	CP	20	45	45	45	45	89	93		
	SP	71-	142-	284-	568-	576-	1136-	1160-		
	Total*	76	164	306	590	598	1178	1206		
PP2	CP	20	42	44	46	50	87	89	175	177
	SP	70+	141+	283+	567+	575+	1135+	1159+	2271+	2319+
	Total*	75	158	302	588	600	1167	1193	2307	2357
PP3	CP	45	42	43	43	45	87	89		
	SP	35+	71-	142-	284-	288-	568-	580-		
	Total*	40	90	162	304	310	608	622		
PP4	CP	20	43	44	46	48	90	92	176	178
	SP	35	70+	141+	283+	287+	567+	579+	1135+	1159+
	Total*	39	88	160	304	310	600	614	1170	1196
PP5	CP	19	42	45	46	46	90	92		
	SP	17++	35+	71-	142-	144-	284-	290-		
	Total*	22	52	89	161	163	315	323		
PP6	CP						88	90	176	180
	SP						283+	289+	567+	579+
	Total*						314	322	598	614
PP7	CP		45	50	53	58	88	91	180	182
	SP		17++	35+	71-	72-	142-	145-	284-	290-
	Total*		40	60	97	103	171	177	317	325
PP8	CP				47	52	86	89	175	181
	SP				71-	72-	142-	145-	284-	290-
	Total*				103	109	181	187	332	344

*Note that some of the pilots fall on the same carriers and therefore they don't
add up as expected.

Appendix D

DVB-T2 Pilot Patterns

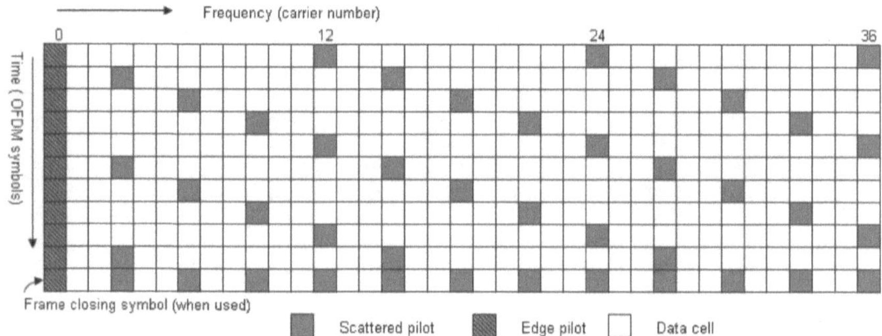

Figure D.1: Scattered pilot pattern PP1 (SISO).

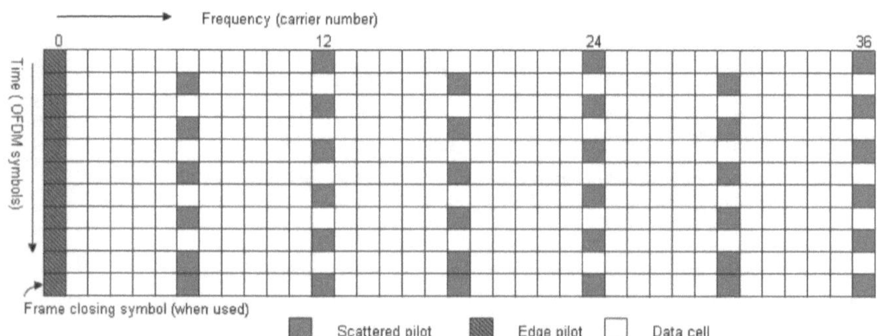

Figure D.2: Scattered pilot pattern PP2 (SISO).

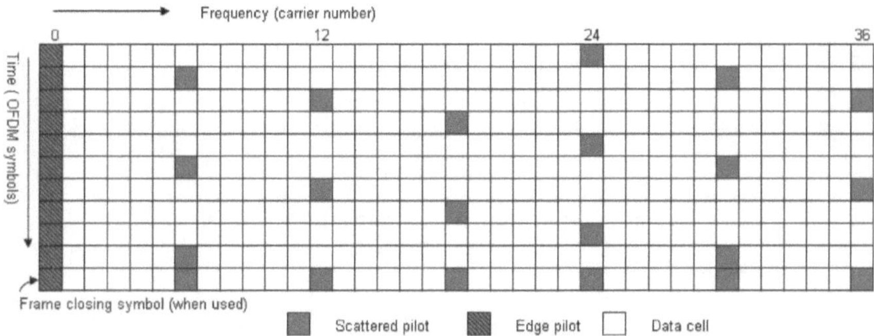

Figure D.3: Scattered pilot pattern PP3 (SISO).

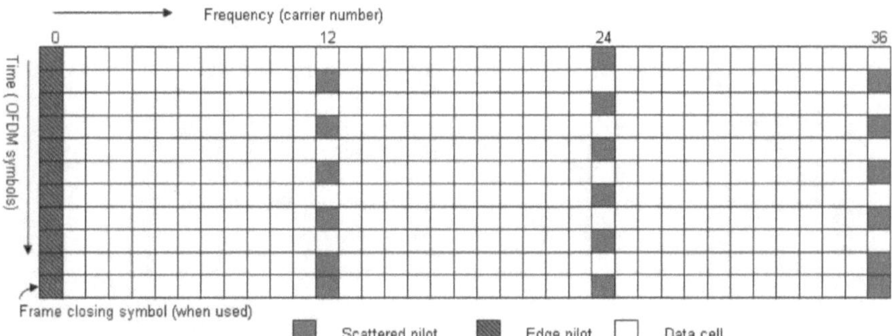

Figure D.4: Scattered pilot pattern PP4 (SISO).

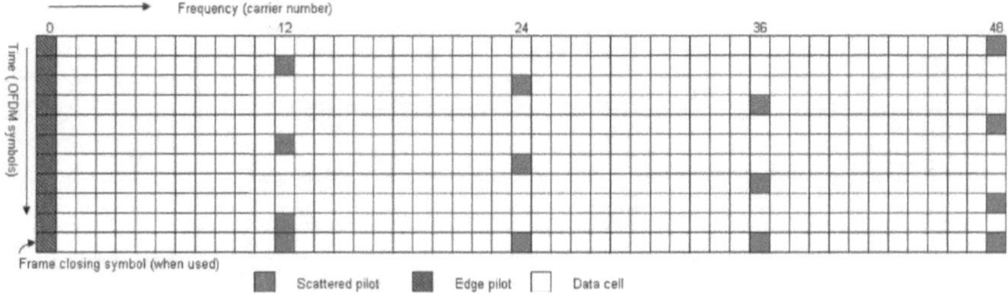

Figure D.5: Scattered pilot pattern PP5 (SISO).

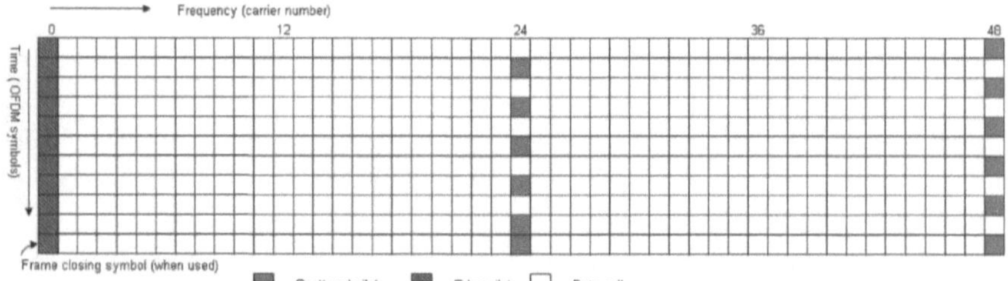

Figure D.6: Scattered pilot pattern PP6 (SISO).

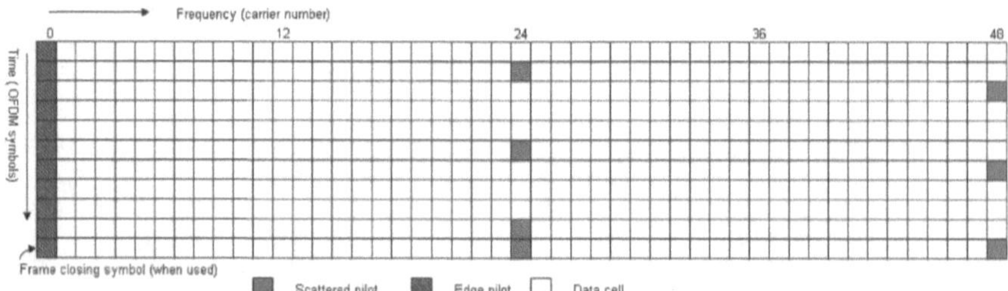

Figure D.7: Scattered pilot pattern PP7 (SISO).

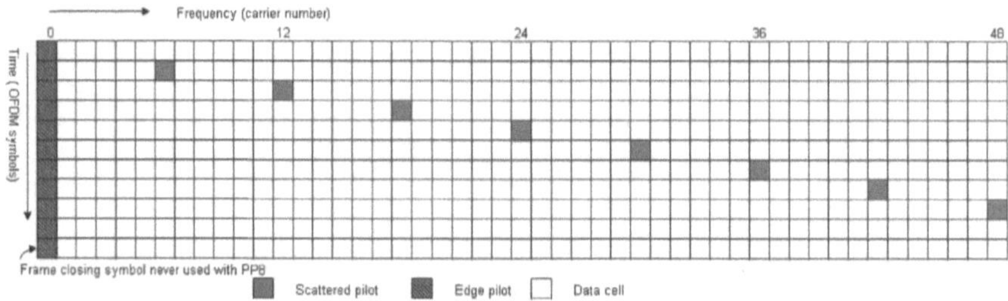

Figure D.8: Scattered pilot pattern PP8 (SISO).

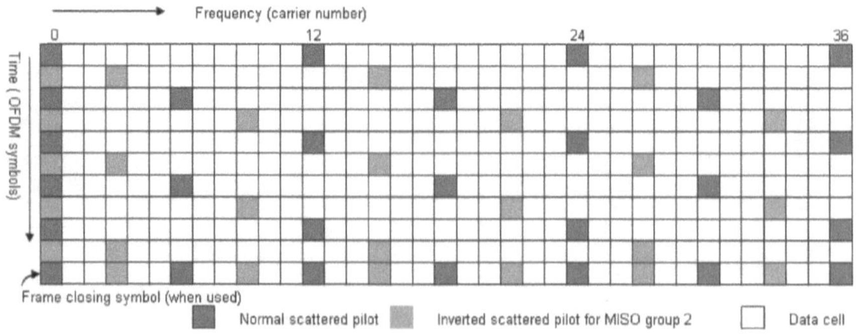

Figure D.9: Scattered pilot pattern PP1 (MISO).

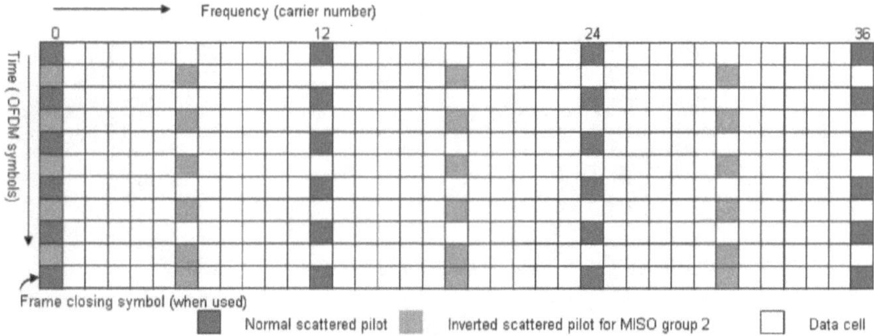

Figure D.10: Scattered pilot pattern PP2 (MISO).

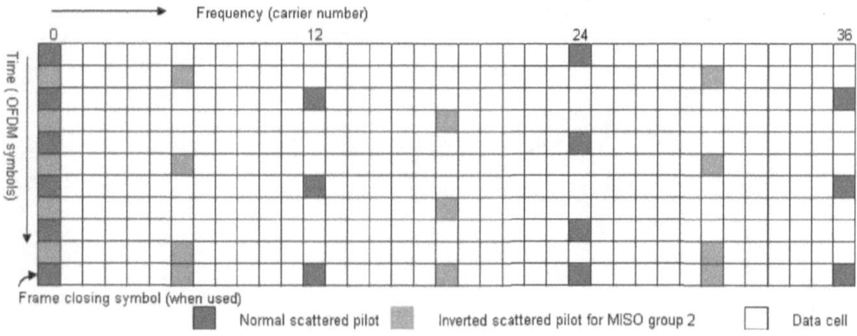

Figure D.11: Scattered pilot pattern PP3 (MISO).

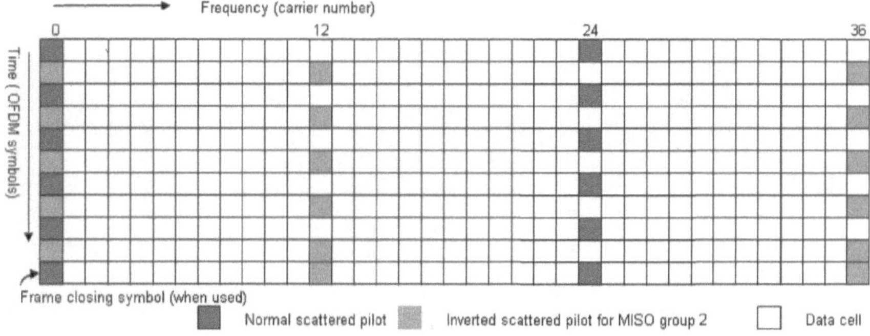

Figure D.12: Scattered pilot pattern PP4 (MISO).

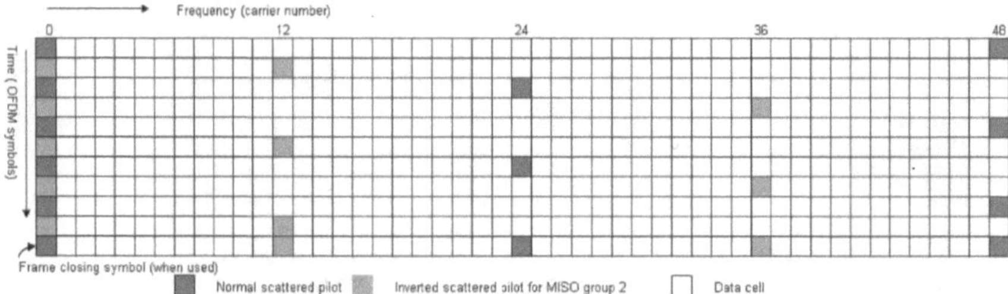

Figure D.13: Scattered pilot pattern PP5 (MISO).

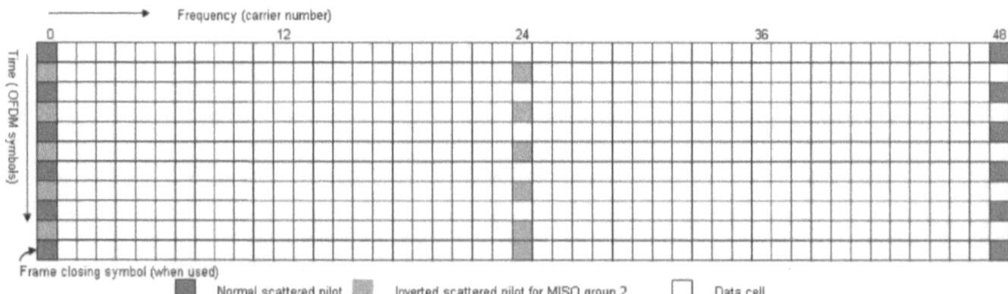

Figure D.14: Scattered pilot pattern PP6 (MISO).

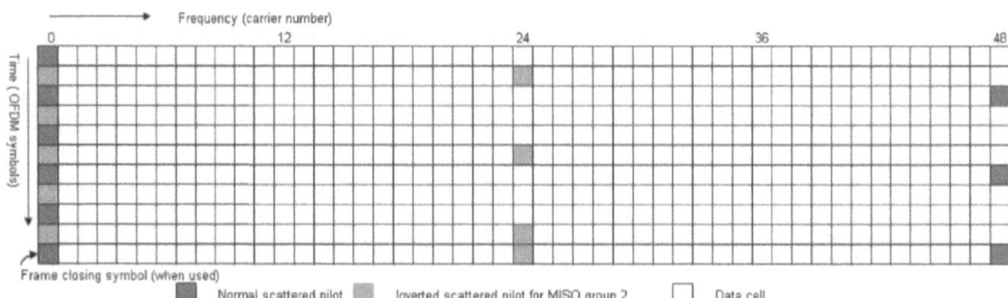

Figure D.15: Scattered pilot pattern PP7 (MISO).

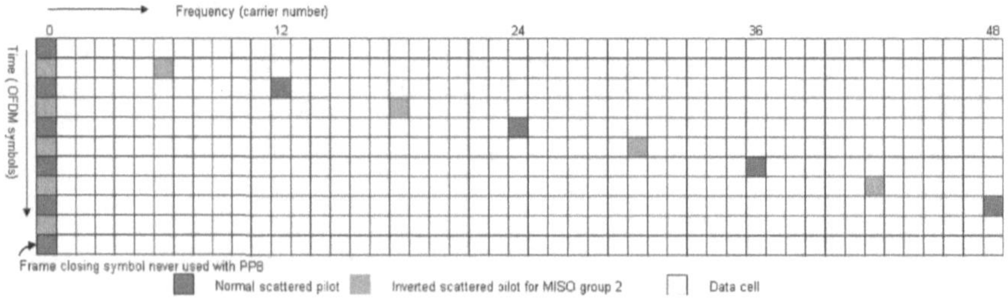

Figure D.16: Scattered pilot pattern PP8 (MISO).